ZERO TO BIRTH

Zero
to
Birth

HOW THE HUMAN BRAIN IS BUILT

W. A. HARRIS

PRINCETON UNIVERSITY PRESS

PRINCETON & OXFORD

Published by Princeton University Press
41 William Street, Princeton, New Jersey 08540
99 Banbury Road, Oxford OX2 6JX

press.princeton.edu

First paperback printing, 2024
Paperback ISBN 978-0-691-25394-7

The Library of Congress has cataloged the cloth edition of this book as follows:

Names: Harris, William A. (William Anthony), author.
Title: Zero to birth : how the human brain is built / W.A. Harris.
Description: Princeton, N.J. : Princeton University Press, [2022] |
    Includes bibliographical references and index.
Identifiers: LCCN 2021048088 (print) | LCCN 2021048089 (ebook) |
    ISBN 9780691211312 (hardback) | ISBN 9780691237077 (ebook)
Subjects: MESH: Brain—growth & development | Brain—embryology |
    Neuronal Plasticity—physiology | BISAC: SCIENCE / Life Sciences /
    Neuroscience | SCIENCE / Life Sciences / Developmental Biology
Classification: LCC QP376 (print) | LCC QP376 (ebook) | NLM WL 300 |
    DDC 612.8/2—dc23
LC record available at https://lccn.loc.gov/2021048088
LC ebook record available at https://lccn.loc.gov/2021048089

Editorial: Ingrid Gnerlich, Whitney Rauenhorst
Cover Design: Michel Vrana
Production: Danielle Amatucci
Publicity: Sara Henning-Stout (US), Kate Farquhar-Thomson (UK)
Cover Credit: 1) Brain map: Hendra Su / iStock, 2) Fetus: Vectortatu / iStock

This book has been composed in Arno

This book is dedicated to all the scientists who have searched for the origins of the human brain and by doing so have contributed to this story.

CONTENTS

ILLUSTRATIONS

PREFACE

A MOTHER PLAYS with the tiny fingers of her new baby. Staring into the child's shining eyes, she says: "Oh, little baby, you are just so amazing!" The baby is of her body yet is completely unique. She not only possesses her own fingers and bright eyes, but she also has a mind of her own. Well protected inside her skull, where the mother cannot see and marvel yet further, the baby's brain is truly a wonder—a living, humming supercomputer with billions of electrically active, squidgy cells called "neurons," with trillions of adjustable connections. Her brain is born ready to gather and store relevant information about the world and to take this information into account when making all sorts of decisions, even when the baby still needs constant parental care. Her brain has the power to synthesize instinct and experience, to be curious and investigate the mysterious, to experiment and invent, to feel new textures and new emotions, and more. In addition, her brain has the ultimate power of being totally integral to her budding sense of self. Emily Dickinson's 1862 poem about the brain says it all much more simply:

> The Brain—is wider than the Sky—
> For—put them side by side—
> The one the other will contain
> With ease—and You—beside—[1]

The brain and how it works has been an object of fascination and inquiry for centuries. And though we have discovered much about it, this organ still holds a multitude of secrets.

One of the human brain's greatest secrets has to do with how it forms in the first place—that is, how it is made and develops in utero, from the point of fertilization to birth. When I first entered the field as a young scientist in the mid-1970s, very little was known about this subject, and part of the reason is that this dynamic process is obviously not something that is very easy to observe. But since then, this field of investigation—developmental neurobiology—has flourished. Scientists around the world have become involved in searching for clues to the origins and formation of the brain, using ever more sophisticated methods, scouring the borderlands of developmental biology, evolutionary biology, genetics, and neuroscience. As a result, the past few decades have witnessed many new and exciting discoveries related to the development and evolution of the brain. In this book, we will explore those discoveries, which have brought us a clearer understanding of how a baby's brain is built. While describing the most exciting and revolutionary experiments that have opened our eyes to the mechanisms involved in brain development, I also describe the questions that motivated those experiments. In addition, I explain how the experiments were done. What were the outcomes and the interpretations—and when were the outcomes surprising and the interpretations wrong? How did these experiments and the knowledge gained from them change our view of the brain and its development? This book is a chronicle of the science that brought us to our current understanding of how a human brain is built.

It does not give away much of the main plot if I tell you that this story starts with a single fertilized egg. This egg generates an embryo, in which a group of cells become committed or assigned to make the brain. The story then follows these cells as they generate the neurons that wire up the growing brain and concludes, as you might expect, with a fully formed human brain. Much of the story happens at the cellular level, as it is at this scale that one can more easily appreciate the steps and basic principles involved in the process of building the brain. From a certain perspective, the tale I tell may read a bit like the biography of a neuron's coming of age. Along the way, some key questions arise: What events lead a certain population of embryonic cells down the pathway to become the cells in our brains? How many kinds of cells are there in the brain? What influence does a neuron's ancestry and its environment have on its specific fate as a type of neuron and as an individual in its own right? How do neurons grow the thread-like extensions called "axons" and "dendrites" that make all the right kinds of electrically active connections, so the brain wires up correctly? Why do so many neurons naturally die in our first years of life? What does a neuron have to go through to become a permanent part of your brain? How will this neuron change over the years? Although it unfolds on the vanishingly small cellular and molecular scales, the biography of a neuron is full of drama, and this book will open a window onto that stage.

Interwoven with the story of brain development is a parallel tale that extends further back in time—a story of human brain evolution more generally, or, in other words, a story of how human beings got human brains. This story also starts with single cells, though these cells existed billions of years ago in the primordial ooze of early life on earth. The evolutionary tale ends at the same place that the developmental story ends: at a fully formed human brain. And so these are not independent threads. Work in the field of evo-devo (evolution and development), for example, has found that many genes and molecular pathways that affect steps in the embryonic development of our brains are the same genes and molecular pathways that drive steps in the evolution of the human species.[2] So, even though this book is largely focused on the developmental story of the brain, I weave in an evolutionary perspective to provide greater context and deeper insights. Seeing the human brain through the lens of its developmental *and* evolutionary origins gives us a richer view of our innate characteristics as humans.

Our evolutionarily inherited human genomes (our human "nature") hold the basic building plans of the human brain, and our interaction with the environment ("nurture") can affect and guide the building process. Conversely, the influence of the environment is often affected by genetic differences among individuals. Nature affects nurture, and nurture affects nature. Those who think too much about nature vs. nurture often ignore another major player: chance. Random events affect many aspects of brain development. The examples given in this book highlight that genes, environment, and Lady Luck all play roles in making one's brain.

As development neurobiologists have learned more about the many genes and molecular pathways that are involved in making the brain, we have found new connections between these genes and a growing number of recognized neurological, psychological, and psychiatric syndromes, some of which arise beyond infancy and childhood. Because thousands of genes and thousands of steps are involved in making the brain, many things can go wrong. To develop strategies to treat these various syndromes, it is extremely valuable to know which genes and molecular pathways are involved in building the brain and ultimately may be involved in the potential treatment of the syndromes in question. Another key medical challenge involves discovering ways of repairing the brain after injury or disease. In the human fetus, billions of neurons are made, but an adult human brain loses the youthful brain's ability to make new neurons to replace any that are lost to injury or disease. Similarly, during development, the neurons of the brain wire up correctly—but in adults, severed axons in the brain are not able to regrow and wire up properly. Because the maturing brain loses the capacity to regenerate new neurons and connections as it ages, fixing "broken" human brains is one of the biggest grails of medical science. It is incredibly challenging not only because building the brain is such a complex and sensitive process (as we will see), but also because so many of the key processes happen in utero. The womb guards its secrets well, and particularly in the realm of experimental science, we must take great care as we seek to make progress. Yet exceptional progress has been and is being made. For example, lessons learned from the study of neural development now allow researchers to direct stem cells in culture to become specific neuron types in the brain, or to become human brain organoids—microscopic human mini-brains—which can be used to study brain diseases and search for cures and brain-repair strategies that can alleviate suffering.

One may wonder: Will knowing more about how the human brain is built help us to better understand how it works? After all, we can learn how to build something without knowing what that something does. However, in this case, we are already armed with a huge and rapidly growing amount of science on how the brain works, and so we already do have a good sense of what the brain does. In this context, learning how to build it from the ground up should help us see *how* it does what it does, and specifically, how information comes to flow efficiently through the brain.

The final chapter in this book focuses on how the human brain evolved from our closest primate relatives and hominid ancestors. What, if anything, is fundamentally different between human brains and the brains of our extinct ancestors and living relatives—and how did these differences arise? What are the special mechanisms that are essential to building a typical, modern human brain, as opposed to the brain of another species? Whatever human brain we were born with changes as a result of experience (especially during childhood), and the storage of personal information, skills, and memories alters the brain. It turns out that the development mechanisms that make human brains ensure that no two human brains are ever the same. What makes us all human also makes us all different.

And so it is a rich and complex story I recount in the pages that follow. From the perspective of an experimental neuroscientist in a field in which I have personally witnessed and taken part, I describe how science has revealed the structures and mechanisms of the developing brain from its earliest embryonic origin to birth and a little beyond. The story progresses chronologically, step by step, tracking the actual growth and development of the human brain. Throughout, findings from studies of a variety of model organisms—such as nematodes, flies, frogs, fish, birds, mice, and sometimes non-human primates—are woven into the narrative, which provides a perspective on the evolutionary process that parallels the developmental one. The book concludes with a discussion of what makes individual brains unique and how research on early neural development is helping us better understand the genetic and embryonic origins of many neurological and cognitive traits that only reveal themselves later in life. The story of how the human brain develops, from conception to birth, is a tale of becoming like no other. It is one that we are continuing to unravel and explore.

ZERO TO BIRTH

1

Rise of the Neurons

In which some embryonic cells become neural
stem cells, the founders of the nervous system,
and in which we get the first glimpses of
the evolution of the brain.

Totipotent Stem Cells

The end of the nineteenth century was a time of tremendous
progress in embryology. Questions that had been debated for
centuries concerning how an organism with all its parts emerges
from a single-cell egg were beginning to be answered by experiments rather than debates. One of the most fundamental of
these questions was: When a fertilized egg cell divides to make
two cells, does each of the two cells have the capability to make
a complete being, or do the two cells divide this potential in
some way? This was a question that just could never be answered by debate. An experiment on real embryos was clearly
necessary to resolve the issue.

In 1888, Wilhelm Roux, working at the Institute for Embryology in Wrocław, took up the challenge of answering this question by using frog embryos at the two-cell stage. He inserted a

heated needle into one of the two cells and then let the embryo develop from the remaining live cell. Most of the experimental embryos ended up looking like halves of animals, for example, a right or left half of an embryo rather than a whole one. Based on these results, Roux argued that the capacity to make a whole animal is indeed divided in two at the very first cell division.[1] As Roux's was the first scientific experiment ever to be done on any type of embryo, he is credited with being the father of the entire field of experimental embryology, which has been a cornerstone of developmental biology ever since.

Roux's results were unimpeachable, but his basic interpretation of them drew immediate concern because it also seemed possible that the dead cell might have affected the development of the single surviving cell next to it. So, a few years later, another embryologist, Hans Driesch, working at a marine biological station in Naples, did a very similar experiment, though he used sea urchin embryos rather than frog embryos. The wonderful thing about the sea urchin embryos is that at the two-cell stage, all it takes is gentle shaking to separate them into single cells. So, in principle, there should be no effects from any neighboring dead cells. The results from Driesch's experiment were the opposite of Roux's. Instead of making half animals, each of the two cells gave rise to an entire sea urchin.[2]

Of course, Driesch's results strengthened suspicions that the presence of the dead cell in Roux's experiments might have affected his results. But it was also plausible that the discrepancy pointed to a fundamental difference in the way that sea urchins and frogs develop. Therefore, it became of major interest to know what would happen if the first two cells of a frog embryo could be fully separated and both cells kept alive. But this experiment was (and still is) extremely challenging because the cells are fused and share their contents at these stages in amphibian embryos. Nevertheless, in 1903, Hans Spemann of the University of Würzburg managed to succeed in doing so by fashioning a tiny noose from a fine hair of his newborn baby's head. He positioned the noose between the two cells and began, ever so slowly, tightening it, little by little, minute by minute, with amazing steadiness of hand. When the noose was fully tightened, the two cells fell apart from each other, both alive. In many instances, both these cells formed a whole embryo.[3] It seems that Roux's interpretation of divided potency was indeed wrong and was probably an artefact of the effects of the dead cell, though the biological reason for Roux's results has never really been further investigated.

What about mammals? In 1959, Andrzej Tarakowski at the University of Warsaw separated single cells from a two- or four-cell mouse embryo and then placed each of them into the wombs of foster mothers. These isolated cells often gave rise to healthy baby mice.[4] Similar experiments have now been done with many other mammals. In humans, identical twins result from a single embryo spontaneously splitting into two, and though it is still not known exactly when or how this splitting occurs, the embryonic cells at the time of such splitting are able to make entire humans. Genetic testing of early human embryos that are fertilized in vitro (IVF embryos) is offered to couples who are at risk of carrying severe genetic abnormalities. In such a procedure, one cell of a human embryo at the four- or the eight-cell stage is removed for testing. If no obvious genetic defects are found, the remaining three- or seven-cell embryo can be reimplanted into the womb, as there is little risk that the removal of just one cell has injured the potential of the remaining cells to make an entire human being. So the results are often happy ones. Thus, the embryonic cells at this stage are said to be "totipotent": capable of making it all.

Genesis of the Brain

Written in our genes is an eons-long history of the human brain's evolution. The information there is used to reconstruct an entirely new brain in every single baby. Each of us begins life as a tiny egg, a single cell smaller than a grain of table salt. The cell, like its evolutionary ancestors all the way back to the dawn of cellular life 4 billion years ago, is surrounded by a membrane and contains a nucleus. Inside the nucleus of the egg cell are the instructions for making an entire human being. A sperm cell, carrying its own set of complementary instructions, finds the egg and pushes itself inside. With a copy of the genome from each parent, the fertilized egg starts to divide. First, it makes two cells. Two cells become four, then eight, and so on. Soon there is an embryo composed of thousands of cells. Each of these cells contains a nucleus, and each nucleus has access to the full set of instructions.

Some of the instructions for making the brain came from single-cell organisms of the Proterozoic eon.[5] These protozoans sensed their local environment and responded accordingly. They did not have brains themselves—but they had the makings of brains. Many modern protozoans are excitable and motile; they search for food and mates, they adapt to new situations, they store memories of events, and they make decisions. Modern single-cell creatures, such as paramecia, are relics of this ancient eon that preceded the origin of multicellular animals by at least a billion years. When a paramecium swims into a wall, it reorients and heads off in a new direction. It is the synchronized beating of the thousands of tiny cilia all over its body that propels the paramecium forward. The mechanical stimulus caused by the bump opens calcium channels in the paramecium's cell membrane. An electrical current carried by calcium ions begins to flow through these channels, and this current changes the voltage across the membrane. Other calcium ion channels in the cell's membrane are sensitive to this voltage change, and they open in response. The opening of these voltage-sensitive channels allows even more calcium to flow across the membrane, which changes the membrane voltage further and opens yet more channels. This explosive electric feedback is the essence of a neural impulse of the kind used by the neurons in our brains, except that neurons tend to use sodium ions rather than calcium ions to generate an impulse. What this electrical impulse does for the paramecium is to let calcium ions enter instantly all over the membrane, which leads to the simultaneous disruption of the beating of the cilia of the paramecium, causing it to tumble. When the cell recovers, it is heading in a new direction. The paramecium's channels that are activated by mechanical deformation and those that are activated by voltage are evolutionarily related to the channels found in the neurons of all animals. It seems that many properties that are characteristic of the brain were already encoded in the DNA of our single-cell ancestors. How they got these neural-like properties lies buried even deeper in the early evolution of life on earth.

Protozoans like paramecia have many specialized functions located in distinct compartments of the cell, such as a digestive system, a respiratory system, cilia for motility, a nucleus to carry key information accumulated since the origin of life itself, and an excitable membranous skin capable of making rapid alterations in behavior. Protozoans must do all this, and much more, in a single cell. With the rise of multicellular animals, cells could specialize and divide the labor. A brain is a collection of neurons that communicate with one another using synapses. Nervous systems with real neurons and synapses did not arise, and could not have arisen, until multicellular life began. Jellyfish are members of a phylum of animals called the cnidarians that arose around 600 million years ago. Cnidarians have networks of interconnected neurons that share many characteristics with the neurons of the bilaterally symmetric animals (aka bilaterians) like us. Bilaterians also arose at one of the earliest of branch points on the tree of multicellular animal life. Cnidarians and bilaterians may have evolved neurons and synapses independently, but it is equally likely that these attributes evolved once in a common ancestor to both groups. The first vertebrate animals arose more than 450 million years ago. These early vertebrates are most related to today's lamprey eels. Lampreys not only have neurons like ours, but they also have a similar layout of the nervous system, including a brain with the anatomical and functional beginnings of the cerebral cortex, the region of the brain that is so greatly expanded in humans.[6]

Finding the Neural Stem Cells

When, where, and how do neurons first arise in an animal? About 3.5 billion years ago, single-cell organisms were sometimes joining together to become simple multicellular life forms, which could then afford to divide tasks among themselves. In the multicellular life form known as a human, cells also begin to take on specific tasks. Some will build muscle and bones, some will make skin, some will make the digestive system, and so on. Those that will make the brain and the rest of the nervous system are the neural stem cells.

If you take a trip to a pond in the woods in early spring and collect some freshly laid frog eggs, one of the first things you might notice about these eggs is that they have a darker half and lighter half (figure 1.1). The darker half is known as the "animal" side, and the lighter half is known is the "vegetal" side. The

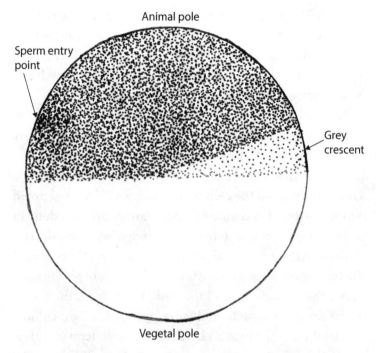

FIGURE 1.1. A frog egg shortly after fertilization. A remnant of the sperm entry point is seen in the aggregation of pigment granules there (near top of figure). The animal pole is at the top, and the vegetal pole is at the bottom. The gray crescent forms opposite the sperm entry point in the animal hemisphere near the equator. The gray crescent marks the dorsal or back side of the developing frog embryo.

imaginary line from the animal pole to the vegetal pole forms the animal-to-vegetal axis of the embryo. When a sperm fertilizes a frog egg, it initiates a movement of the dark pigment granules toward the point of sperm entry. This movement leads to a lightening on the opposite side of the egg, where one can see what is known as the "gray crescent" rising like a new moon. The gray crescent is on the side of the frog embryo that will become the dorsal or back side of the future tadpole. We can now draw another imaginary line from dorsal to ventral (back to belly). These dark, light, and gray landmarks remain until the frog embryo reaches a stage of development known as the blastula. The blastula is basically a ball of several hundred cells with a fluid-filled hollow in the middle. Human embryos reach this blastula stage about one week after fertilization.

Embryologists of the late 1800s wanted to understand how this ball of cells transformed itself into a little tadpole, so they began to follow cells that were consistently positioned at certain coordinates along animal-to-vegetal and dorsal-to-ventral axes. They stained the cells with permanent dyes and noted where the dye ended up. Such experiments are now done in embryology courses at universities throughout the world, and students in these courses discover for themselves the origins of the three great germ layers of the vertebrate embryo: the ectoderm, the mesoderm, and the endoderm (from the Greek words for outer, middle, and inner layers). The light-colored vegetal third of the blastula becomes the endoderm and gives rise to the digestive tract and its organ systems. The equatorial third between animal and vegetal poles, which contains the gray crescent on its dorsal side, becomes the mesoderm, which gives rise to muscles and bones. The dark animal portion of the embryo, known as the animal cap, becomes the ectoderm, giving rise to the epidermis and the nervous system. Students in such embryology courses often go further and find that the primordial nervous system comes from just the dorsal half of the ectoderm, the region that lies directly above the gray crescent.

The Organizer

Knowing which cells of the blastula will become the neural stem cells allowed Hans Spemann, now working in Freiburg, to devise an experiment to test whether these cells are also capable of giving rise to other tissues or whether they have become re-stricted to making only the nervous system. Spemann thought of testing this by taking groups of cells from a particular position on one embryo and transplanting them to a different position on another embryo. As was his style, Spemann invented a vari-ety of new microtools for these experiments, including incred-ible fine-glass pipettes with fingertip control that could be used to transfer tiny fragments of embryonic tissue carefully between embryos, and superfine scalpels to cut out such fragments. With such tools and his extreme dexterity, Spemann was able to perform precise cut-and-paste experiments on amphibian embryos. In one series of experiments, he transplanted bits of one blastula to different positions on another. When he trans-planted a piece of the dorsal ectoderm from the blastula of a newt embryo (i.e., the piece of the embryo that would have become its nervous system if left in its original position) to a different position in the blastula of another newt embryo, nothing ex-traordinary happened. The resulting animal developed nor-mally. It did not, for example, have an extra bit of brain tissue. The transplanted cells simply switched or ignored their previ-ous fates and integrated beautifully into their new positions. They still appeared to be totipotent and flexible at this stage.

The breakthrough came at the next stage of development, just two to three hours later. This is called the "gastrula stage." Human embryos reach this stage at about week three of gesta-tion, when there are thousands of cells. The gastrula stage of development begins when the cells of the mesoderm start to move into the hollow in the center of the blastula. Developmen-tal biologists say that they begin to "involute." Imagine holding a soft balloon in your left hand; now push the fingers of your right hand into the balloon. The first mesodermal cells to invo-lute are the most dorsal ones (figure 1.2). These are the cells of

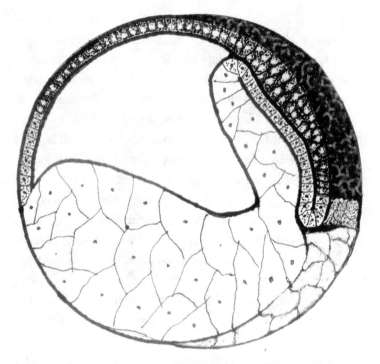

FIGURE 1.2. A cross section of an amphibian embryo during gastrulation and neural induction. The involuting mesoderm (gray stippling) is moving under the dorsal ectoderm (dark) and inducing the latter to become neural ectoderm, which can be seen thickening up as the neuroepithelium.

the gray crescent. When Spemann transplanted just a small piece of this involuting dorsal mesoderm at the very beginning of gastrulation from one donor newt embryo to the ventral side of another host embryo, something remarkable happened. Spemann was stunned! The host animal did not look normal, as happened when he did this experiment at the blastula stage. Nor did it have an extra bit of out-of-place mesoderm, as one might have suspected if the transplanted tissue had become restricted. What Spemann saw was that a whole new secondary embryo developed in these hosts.[7] This second embryo was often joined belly-to-belly with the host embryo, like face-to-face Siamese twins!

What happens during gastrulation is absolutely critical for the organization of an embryo. Without gastrulation, any frog or even any human embryo would not have much of a body and no brain at all. This is why Lewis Wolpert, a British developmental biologist who will be discussed in the next chapter, often told his audiences at lectures: "It is not birth, marriage, or death, but gastrulation which is truly the most important time in your life." How to explain this incredible result in terms of cells, tissues, and biological mechanisms was Spemann's next challenge. There were two main possibilities. One was that the transplanted piece of dorsal mesoderm was still totipotent, and that the trauma of being transplanted somehow stimulated these cells to make an entire new embryo. The other possibility was that the transplanted tissue somehow induced the nearby host tissue to form the new embryo around it.

Spemann had a brilliant young graduate student, Hilde Proescholdt, who took up the challenge of disentangling these possibilities as her thesis project. It was clear that if the transplanted dorsal mesoderm grew into one of the twins by itself, then this twin would be composed of donor-derived cells. However, if the transplant somehow induced the surrounding tissue to make an embryo, then this second embryo would be composed mostly of host-derived cells. So, Proescholdt addressed this issue by using embryos of two species of newts, one that was lightly pigmented (which she used as the donors) and one that was darkly pigmented (which she used as the hosts). The cells of the light embryos could be identified in a microscope from their lack of pigment granules. Then, just as Spemann had

done, she transplanted this special piece of the dorsal mesoderm from one early gastrula to the ventral side of another—the
only difference being that this time, the donor cells were light,
and the host cells were dark.

Her experiments immediately settled the issue. She found
that the transplanted cells made only a minor contribution to
the second embryo (figure 1.3). Most of the second embryo,
including the brain and spinal cord, was made of host rather
than donor cells.[8] With this one experiment, she proved that
this small piece of dorsal mesoderm, taken at the beginning of
gastrulation, can induce the tissue around it to make an entire
embryo. Spemann said it in this way: "This experiment shows,
therefore, that there is an area in the embryo whose parts,
when transplanted into an indifferent part of another embryo,
there organize the primordia for a secondary embryo."[9] Spemann called this tissue "the organizer." The discovery of the
organizer is one of the most fundamental findings in all developmental biology.

After writing up her PhD thesis on this work, Proescholdt
married Otto Mangold and moved with her husband and their
new baby to Berlin. Tragically, soon after the move, a gas heater
exploded in their new home. She suffered horrific burns and did
not survive to see the publication of her famous thesis in 1924
nor the award of the Nobel Prize to Hans Spemann in 1935 for
their joint discovery of the organizer.

The organizer region of a frog embryo is similar to what is
known in a mammal embryo as the "node." The mammalian
node, like Spemann's organizer, is a region of dorsal mesoderm
that involutes and induces the overlying ectoderm to make
neural stem cells. The node or organizer region must work in
a similar way in all vertebrate animals, as the node from a chick

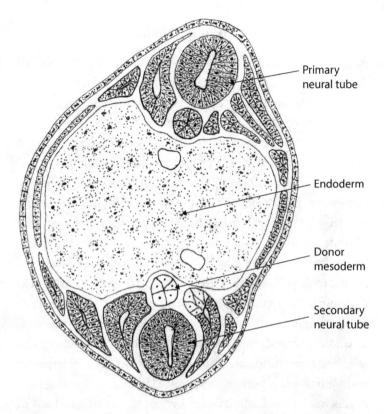

Primary
neural tube

Endoderm

Donor
mesoderm

Secondary
neural tube

FIGURE 1.3. A result from Spemann and Mangold's 1924 experiment. Hilde Mangold (née Proescholdt) made cross sections of the pigmented newt embryos that had organizer transplants from unpigmented donor embryos. What she often saw, as shown here, was the unpigmented donor mesoderm underneath the host-derived secondary neural tube.

embryo can act like an organizer when transplanted into a frog embryo, and the node from a mouse embryo can induce a secondary chick embryo. Similar results have now been found with mouse-to-frog, chick-to-fish, fish-to-frog, chick-to-mouse, and mouse-to-chick transplants.

The Neural Inducer

As soon as Mangold (née Proescholdt) and Spemann published their findings, biologists immediately wanted to know how the organizer worked. How can a small piece of tissue orchestrate the building of an entire embryo around it? How does the organizer communicate with neighboring cells, and what does it tell them? Does it, for instance, tell some of them to make the brain? Such questions became a major preoccupation of developmental biology laboratories around the world. It was quickly discovered that the organizer tissue did not have to heal into place and involute to induce a second embryo; one could just stuff it into the hollow center of a blastula, and it was still able to induce a second embryo from the surrounding host tissue. It even worked if the organizer tissue was separated from host tissue by a piece of filter paper, so direct cell-to-cell contact was not essential. These experiments made it seem likely that the organizer was releasing some diffusible signaling molecules. The interspecies node-transplant experiments suggested that these signaling molecules were a fundamental and ancient aspect of how balls of cells become organized embryos, so there was great interest in discovering the nature of these magic molecules.

The host cells that were closest to the transplanted organizer generally became the central nervous system of the secondary embryo, so the search for the organizer substance became, in some laboratories, the search for the "neural inducer," a hypothetical substance released by the organizer that was responsible for turning totipotent cells of the blastula into the neural stem cells of the gastrula.

Some laboratories tried to find organizer substances or neural inducers through biochemical analysis of organizer tissue, but the miniscule amount of starting material stifled progress using this approach. Other laboratories searched for other tissues that might have organizer properties; they found that bits of liver and kidney were capable of acting like the organizer if they were stuffed inside a blastula. But after a while, it became clear that just too many different tissues had neural-inducing capabilities. In 1955, a disheartened Johannes Holtfreter, one of those on the hunt for the neural inducer, said in despair that "fragments from practically every organ and tissue from various amphibians, reptiles, birds and mammals, including man, were inductive."[10] Even random chemicals off the laboratory shelf were sometimes inductive. It seemed the problem was that the animal cap cells of newt embryos were somehow poised to become neural, so finding the thing that normally induced them was going to be a huge challenge. As a result, the hunt for the real neural inducer went cold for decades.

A small digression is now warranted. In 1927, a British endocrinologist, Lancelot Hogben, relocated to rural South Africa and found himself surrounded by multitudes of claw-toed frogs, known as *Xenopus*. Hogben immediately took advantage of their abundance for his research on hormones. He injected adult female *Xenopus* with an extract from the pituitary gland of an ox, and to his astonishment saw that the injected frogs soon started laying an abundance of eggs. Hogben knew that the urine of pregnant women also carried some pituitary hormones, so he and his colleagues tested the effects of injecting concentrated urine from potentially expectant mothers into adult female *Xenopus* and found that egg-laying predicted pregnancy very accurately. As a result, *Xenopus* became used for pregnancy tests throughout the world until the 1960s.

More important to the field of developmental biology was the fact that one could get *Xenopus* eggs on demand throughout the year, just by injecting females with hormones, rather than seasonally, as was the case for newts and salamanders. Early in my own career, I worked with salamander embryos, so my embryological experiments were restricted to springtime. I must say, I liked the seasonal pace of the work. Later, I switched to *Xenopus* embryos because they were so much more readily available, and work could proceed faster. The luckiest thing about *Xenopus*, however, for those who were still searching for the neural inducer, was that the animal cap cells of *Xenopus* offered a clean experimental system for a new molecular approach to searching for the organizer. If one cuts out the animal cap of a *Xenopus* embryo and puts it in a petri dish, it does not make any neural tissue, unlike the case for tissue from newts and salamanders, where even this small insult is enough to do so. The *Xenopus* animal cap, when isolated in a petri dish, becomes pure epidermis. If, however, one waits a couple hours until gastrulation is in progress and then puts these same animal caps in a petri dish, they make neural tissue. This clear change in the commitment from epidermal to neural tissue that can be seen in isolated *Xenopus* animal caps offered a new way to search for the elusive neural inducer.

Sixty-eight years passed between the first report of the organizer by Mangold and Spemann and the moment in 1992, when Richard Harland and his group at the University of California, Berkeley, taking advantage of *Xenopus* embryos and modern molecular biological strategies, announced the discovery of the first active component of Spemann's organizer.[11] It was a neural inducer. Harland and colleagues called the protein they had discovered "Noggin," which is slang for "head." Noggin is made and secreted by cells of the organizer, and it is able to directly induce totipotent embryonic stem cells to become neural stem cells.

The Secret of Neural Induction and Growing Human Mini-brains

Most developmental neurobiologists, including me, assumed that when neural inducers were eventually discovered, they would turn out to be molecules that instruct cells to become neural stem cells. So, we figured, this was probably what Noggin was doing. But this assumption was wrong. This kind of thing often happens in biology. You are biased to suspect that something works one way, but it turns out that it works in almost exactly the opposite way. So it was for neural induction. The first part of this reversal of general expectations came from Doug Melton's laboratory in the Department of Biochemistry and Molecular Biology at Harvard University. Melton was searching for a signaling protein that, when applied to animal caps of *Xenopus* embryos, turned them into mesodermal tissue: muscle and bone. They had narrowed down their search to a class of signaling proteins. A postdoc in Melton's lab, Ali Hemmati-Brivanlou, found a way block the reception of this potential mesoderm-inducing signal. As he and Melton hoped, the animal caps of embryos that were treated in this way did not become mesoderm even when exposed to the mesoderm-inducing signal. But the thing that came as a surprise to everyone was that these animal caps became neural just as they did when they were exposed to neural inducers like Noggin.[12]

This new result raised what seemed like a shocking possibility: Noggin might *not* be instructive; it might *not* induce cells to become neural. Instead, it might simply stop them from becoming something else. Indeed, this turned out to be the case. There is a signal that percolates through the animal cap telling cells to become epidermal. Noggin works by blocking this signal.

Noggin is not instructive; it does not tell cells to become neural stem cells; it simply prevents them from becoming epidermal. So the simple secret of "neural induction" is that the term "induction" is completely inappropriate, because inducing cells to become neural is exactly what the neural inducer does not do. The cells will become neural stem cells by default as long as the "neural inducer" prevents them from being induced to become epidermal.

Neural inducers like Noggin (several others were subsequently found) are now known to work by blocking a set of signaling molecules known as bone morphogenetic proteins (BMPs).[13] BMPs are secreted proteins that induce ectodermal cells to become epidermal. BMPs were named for their ability to induce the formation of bone, but they have since been found to have effects throughout the body, especially in early development. The mechanism by which Noggin and other neural inducers block BMP signaling is simple. They disguise themselves as receptor molecules for BMPs, and they sponge up all the BMPs that are floating around nearby, thereby preventing BMPs from finding their true receptors. Cells that are not in the vicinity of the organizer, however, are not protected by these BMP sponges, and so they receive a dose of BMP signal that results in their turning on genes that commit them to an epidermal fate. Epidermal cells make even more BMPs and release them onto their neighbors, creating a wave of epidermal induction that spreads across the whole of the animal cap, turning cells into epidermal stem cells. Were it not for the molecules of Noggin and other anti-BMPs protecting some of these cells from being influenced by the spreading wave of BMP, there would be no nervous system, no brain. Anti-BMPs like Noggin are released from the nodes of bird and mammalian embryos, which is why nodes are able to induce neural tissue across species boundaries.

That all vertebrate animals use the same basic molecular mechanisms to generate neural tissue raises the possibility that these mechanisms predate even the origin of vertebrates. At the beginning of the eighteenth century, the French naturalist Étienne Geoffroy Saint-Hilaire emphasized a fundamental similarity among all animals. He noted, as many others had before him, that all animals are composed of essentially the same organs and parts. All animals have digestive systems, circulatory systems, secretory systems, musculoskeletal systems, outer coverings (skin or cuticle), nervous systems, and so forth. The systems may look different in a worm, a fly, a squid, and a human, but they each have all these parts.

A possibly apocryphal story is that, at a dinner party where lobster was served, Saint-Hilaire entertained his dinner guests by observing that the cooked invertebrate animals lying on their backs on the dinner plates looked remarkably like vertebrates in some ways. In a right-side-up lobster, the nervous system is ventral, and the organs of the digestive system are dorsal, opposite to the case in vertebrates. So the upside-down lobsters had the same arrangement of parts as a right-side-up vertebrate. This speculation became known as Saint-Hilaire's inversion hypothesis. The inversion hypothesis was ridiculed and then ignored over the course of the next 150 years. Then, in 1996, a reexamination of the inversion hypothesis was triggered by a study by Ethan Bier, working at the University of California, San Diego. Bier discovered that the fruit fly embryo expresses a BMP dorsally and anti-BMPs ventrally.[14] He showed that blocking BMP signaling ventrally is necessary for the nervous system to form there. It is the same molecular logic as in vertebrates but just inverted, flipped belly-to-back. The resurrection of Saint-Hilaire's inversion hypothesis has led evolutionary biologists to seriously entertain the possibility of a "flip" that led to the origin of vertebrate animals in the Cambrian Period, about a half billion years ago.

In 2012, John Gurdon shared the Nobel prize with Shinya Yamanaka for their work showing how almost any cell in the body could be reprogrammed to become more like a totipotent embryonic stem cell. The ability to reprogram cells to this embryonic state means that we can now clone animals. Gurdon was the first to clone a new animal from the nuclei of an adult.[15] It was a claw-toed frog, a *Xenopus*. Since then, sheep (Dolly), horses, cats, dogs, and monkeys have been cloned. In the futuristic comedy "Sleeper," a botched attempt was made to clone the great leader from some surviving cells from inside his nose. A few years later, workers at Columbia University were able to clone a whole mouse using a reprogrammed olfactory neuron.[16]

There has been huge excitement over the past several decades as developmental biologists have learned more about how to grow totipotent stem cells in tissue culture and how to control the differentiation of these cells, especially into different brain regions. It is now possible to remove a few cells from any human, expose them to a regime of molecular reprogramming so that they become like embryonic stem cells, and then expand these cells in tissue culture, and when there are enough of them, "induce" them to become neural stem cells by exposing them to neural inducers that block BMP signaling. In 2011, Yoshiki Sasai of the Riken Center for Developmental Biology, in Kobe, Japan, found that by using techniques learned from developmental biology, he could induce embryonic stem cells to form layered neural structures, such as the retina and cerebral cortex.[17] Sasai was a hero of mine both for his extraordinary work on the early development of the nervous system and for his breakthroughs in making neural tissues in culture. Thanks in large part to Sasai's work, scientists recognized the huge potential of using such strategies to study human development and disease. Sadly, we lost Sasai, because a postdoc in his laboratory sought instant fame by publishing a simple way to reprogram adult cells by dipping them briefly in an acidic solution. As the postdoc expected, his papers made headlines, but other labs could not reproduce the results, and an internal investigation by the Riken Center found out why they could not: the postdoc had made them up! Though Sasai himself was cleared of having any involvement with the phony data, he was held responsible for a failure of oversight. Sasai was deeply ashamed, he became depressed and committed suicide just six months after the publication of the papers. What a loss! A few years later, the now tried-and-tested biochemical methods that Sasai helped develop are regularly used to reprogram cells in many labs and hospitals. Cells from patients with genetically caused neurological disorders are being used to make microscopic mini-brains that float around in a petri dish. These miniature bits of brain often display similar problems to the patient, speeding medical progress.[18]

Though it is so exciting to be able to make and study mini-brains in a dish, only the neural stem cells inside a human embryo can make an entire human brain. It is the next step in multigenerational stories of these primordial neural stem cells and their descendants that we follow in chapter 2.

2

Bauplan of the Brain

In which the nervous system becomes organized
into discrete regions, and master regulator
genes set evolutionarily ancient patterning
processes in motion.

The Neural Tube

During the third week of gestation, the neural stem cells of a
human embryo are packed tightly together into a thin sheet of
tissue known as the neuroepithelium. The neuroepithelium
covers part of the surface of the embryo. It is attached and continuous at its edges with the epidermal epithelium that sur-
rounds it. Imagine that an orange in your fruit bowl is a newt
embryo that has just finished gastrulation. Inside are the mesoderm and the endoderm. They will make the muscles, bones,
and guts of the animal. The rind of the orange is the ectoderm.
Take a pen and draw a circular shape on the rind. The line you
have drawn is like the boundary between the epidermal stem
cells (outside the circle) and the neural stem cells (inside the
circle). It is a continent of neuroepithelium surrounded by an
ocean of epidermal epithelium. In a human embryo, if one

looks closely at this border between the neuroepithelium and epidermal epithelium, one can see that it rises a little, like a rim, making the neuroepithelium look a bit like a dinner plate. Therefore, at this stage, the sheet of neuroepithelium is called the "neural plate." The neural plate is not as circular as a dinner plate: it is longer from front to back than it is from left to right, and the front part is broader than the rear part.

The cells of the embryo keep dividing, so the embryo is growing. It is getting longer faster than it is getting wider, and as this growth continues, it stretches the embryo and the neural plate in this head-to-tail axis. Meanwhile, the rims at the left and right edges of the neural plate rise higher and begin to curl inward. The rising rims take on the shapes of ocean waves just as they start to roll inward and break. The wave from right side and the one from the left travel toward each other from opposite directions, and their crests begin to collide. It is this image of waves crashing into each other that one can see in the anatomy of embryos at this stage (figure 2.1). When the neural crests do meet, they fuse with each other, transforming the neural plate into the neural tube. The walls of the neural tube are made out of the neuroepithelium. The hollow at the center of the tube is destined to become the central canal of the spinal cord and ventricles of the brain, through which cerebrospinal fluid will soon begin to percolate.

When the neural crests rise, they drag up the attached sheet of epidermal epithelium. So, sheets of epidermal epithelium also meet at the midline, and they also fuse together just above the neural tube. These events enclose the developing nervous system safely inside the outer epidermal layer of the embryo, which begins to make protective skin. By about four weeks of gestation, the human neural tube has usually finished closing.

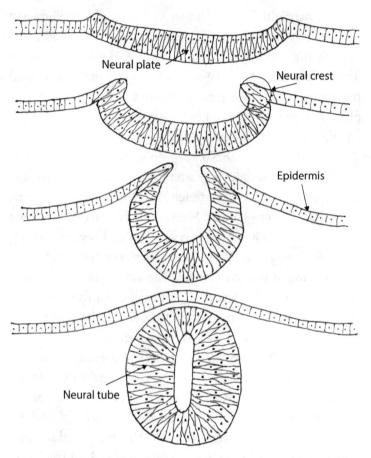

FIGURE 2.1. Closing the neural tube. Shown are cross sections of the neural plate transforming into a neural tube. The neuroepithelium of the neural plate is thicker than the developing epidermis. The neural plate begins to curl up, rising at the edges, known as the neural crests. The neural crests merge with each other at the top (dorsal), creating a neural tube made of neuroepithelial cells. The epidermal cells also fuse together, enclosing the neural tube inside the skin.

The successful closure of the neural tube is a crucial event in every person's life. Defects in neural tube formation are not uncommon, having an incidence of about one in every thousand or so pregnancies. The type of neural tube defect that leads to the failure of the brain to form, known as anencephaly, is completely lethal, whereas *Spina bifida* (from the Latin for "split spine"), the incomplete closure of the neural tube and overlying epidermis, is the most common neural tube defect, accounting for about 40% of all neural tube defects. It is also compatible with postnatal survival if it happens in the spinal region. The gap caused by the incomplete closure of the neural tube in spina bifida may be sutured shut by a surgeon shortly after birth, but usually the development of the spinal cord already has been seriously compromised, leading to lifelong issues with mobility and sensation below the lesion, and neurological problems above the lesion caused by problems in the circulation and maintenance of cerebrospinal fluid, and a shortened average lifespan. This is why some efforts are now being put into developing surgical repair procedures that can be done while the fetus is still in the womb, as correcting the defect as early as possible is likely to lead to better outcomes.[1] The incidence of neural tube defects in humans is reduced by about 70% by making sure that women have sufficient vitamin B9 (aka folate or folic acid) during early pregnancy. However, because the neural tube closes in the embryo before most women even know that they are pregnant, folate supplements taken after this time are ineffective in preventing neural tube defects. That is why some countries, such as Canada and the United States, now have legal requirements to fortify grains, cereals, and flour with folate—which has resulted in a dramatic decrease in the incidence of neural tube defects in these countries. Many thousands of cases of neural tube defects could be prevented each year if other countries followed suit!

The Phylotypic Brain

The neural tube stage is also known as the tailbud stage. Human embryos reach this stage at about one month of gestation, when they are still tiny, not much bigger than a sesame seed. Human embryos at this stage also appear to have little tails. The coccyx, or tailbone, of an adult human is a vestige of several separate tail bones in our vertebrate ancestors, which over the course of evolution fused together into one small bone at the base of the spine. A similar sort of sequence happens during human development, whereby a tail-like structure develops and is then resorbed during development.

The presence of a tail-like structure in human embryos makes them look more like the embryos of other vertebrates that do have tails. Aristotle noted that embryonic animals of different species tend to look more alike than the adults do. Embryos of different species even look more like other embryos than they do adults of their own species. On the basis of many such observations made over centuries, the naturalist Karl von Baer (1792–1876) proposed a first law of embryology, which states that for related species in a group, features that are common to the group develop before those features that are not shared. This was perhaps the first key insight into the relationship between evolution and development, whose stories intertwine. The ambitious embryologist Ernst Haeckel began to search for his own law of embryology. It was he who coined the now rather infamous phrase "ontogeny recapitulates phylogeny." With these three words, Haeckel proposed that during development, every animal goes through a set of development stages, and each of these stages resembles the more adult stages of its evolutionarily successive ancestors, which would, for example, explain why humans at an early stage seem to have little tails.

Haeckel has become a notorious figure in the history of science because he doctored his drawings of embryos to exaggerate the characteristics that supported his hypothesis. This doctoring of images played particularly to his fall from grace because, although provocative, his idea was wrong. As the evolutionary biologist Stephen J. Gould pointed out in his book *Ontogeny and Phylogeny*, evolutionary changes are not just simple add-ons to the adult stages of ancestors. Evolution can and does act on many stages of development.[2] Yet development is a drawn-out process in which animals generally become more and more complex. Thus, evolution has more to play with at later stages of embryogenesis than at earlier stages, which fits von Baer's law of embryology. Haeckel's idea of going through a set of stages resembling adult stages of ancestors now seems ridiculous.

In vertebrates and other groups of animals, it is often the case that the very earliest stages of development are also useful in distinguishing species from one another. Take birds, for example. One only needs to look at the huge white egg of an ostrich to have an excellent idea of what is going to hatch out of it, or to see a clutch of three small speckled sky-blue eggs to guess that robins will hatch from them. Experts can identify hundreds of species of bird's eggs. So, it seems surprising somehow, that at the tailbud stage, all vertebrate embryos (fish, amphibian, reptile, bird, and mammal) look amazingly alike and are difficult to distinguish. The fact that at the earliest and the latest stages of development, different vertebrates look more dissimilar to one another than they do at the tailbud stage has led to the "hourglass" model of evolution and development.[3] In the hourglass model, there is a period—neither the beginning nor end, but somewhere in the middle of development—when the anatomies of embryos in a group are most similar. This period of greatest similarity in the embryos of all the members of the group is known as the phylotypic stage.

The phylotypic stage in vertebrate animals is the tailbud stage, and the neural tube is particularly phylotypic at this stage. The neural tubes of all vertebrates are very elongated from head to tail. They are rather straight or gently curved along the back, narrowing as they reach the tail. The front of the neural tube, the part that will make the brain, is expanded and has some consistent curves and flexures. Along this tube is a set of distinct constrictions, as if tight elastic bands were put at specific places along an elongated modeling balloon, and between these constrictions are bulges of neuroepithelium that are in similar positions in all vertebrates (figure 2.2).

In the 1920s, Nils Holmgren recognized that these anatomical similarities in the embryonic neural tubes of all vertebrate animals might constitute the archetypal Bauplan (from the German for "building plan") for the vertebrate brain. For example, the three expanded bulges at the front of the neural tube make the forebrain, the midbrain, and the hindbrain in all vertebrates. Holmgren's insight led to the discovery of several anatomical similarities among different brain parts in different species that were previously unrecognized.[4] For example, differences between the brains of birds and mammals were previously thought to be radical (i.e., the complete absence of a cerebral cortex in birds). But when looked at through the light of the Bauplan of the brain, one can see in bird embryos that the same region of the neuroepithelium that folds outward in mammals—to give rise to the cerebral cortex, the rind, or neural tissue on the outside of the brain—instead folds inward in the brains of birds. Recent studies show that this inner region of the bird brain is full

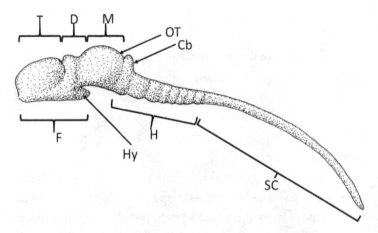

FIGURE 2.2. The vertebrate neural tube at the tailbud stage. At this phylotypic stage, the developing central nervous system looks quite similar in all species of vertebrate animals. The forebrain (F) is composed of the telencephalon (T), the top (dorsal) part of which becomes the cerebral cortex, and the diencephalon (D), the bottom part of which becomes the hypothalamus (Hy). The midbrain (M) is behind the fore-brain, and the dorsal part of the midbrain becomes the optic tectum (OT). Behind the midbrain is the hindbrain (H), which is composed of a series of bulges, the most rostral or anterior of which becomes the cerebellum (Cb). Behind the hindbrain is the spinal cord (SC).

of similar neurons, connected in similar ways to the neurons of the mammalian cerebral cortex.

Cephalization

One of the most basic questions about brains is: Why are they in our heads? The most ancient animals with nervous systems had no head and no brain. The most unchanged descendants of these early animals with nervous systems are the radially symmetric cnidarians, like jellyfish. Jellyfish have a nervous system but no brain. Their neurons are scattered throughout the entire body. There is no central command center. Bilaterally symmetrical animals evolved about 650 million years ago, and they began to move predominantly in one direction, defined as forward. These forward-moving animals shifted their mouths to the front, along with sensory organs and many neurons that process and integrate this sensory information. This is the evolutionary process called "cephalization," by which animals have evolved a head region, in which there is a concentration of neurons called a brain.

The most highly cephalized animals are some insects like bees, the cephalopods like the octopus, and vertebrates like us. Evolutionary trees show that vertebrates share a more recent common ancestor with the less cephalized starfish than we do with an octopus or a bee. Similarly, cephalopods and insects are more closely related to less cephalized groups than they are to each other. This suggests that cephalization is a powerful evolutionary force that operated independently in at least three large branches of the animal kingdom.

In all vertebrate animals, cephalization starts at the time of neural induction, as the front half of the neural plate, which will make the brain, is already much bigger than the back half, which will make the spinal cord. After the neural tube closes, several bulges, flexures, and constrictions begin to appear, and the neural tube starts to divide itself into the discrete regions of the brain: the forebrain, the midbrain, and the hindbrain. As development proceeds, more constrictions and more regions become apparent. The hindbrain becomes divided into several mini-bulges, each of which will give rise to a different part of the hindbrain (e.g., the most forward or rostral of these bulges will generate the cerebellum). The neural tube is becoming segmented—like a worm. This segmentation is obvious in the adult human spine. We have 33 vertebrae, and through each gap between them runs a pair of spinal nerves. Each pair of segmental nerves connects the spinal cord with a specific region of the body. The forebrain and midbrain regions of the neural tube also become segmented. For example, the forebrain becomes divided into an anterior region called the "telencephalon" (Greek for "end" and "brain"), the source of the cerebral cortex and a posterior region called the "diencephalon" (Greek for "in-between" and "brain"), which is the origin of the retina, the thalamus, and the hypothalamus. All the subregions of the neural tube give rise to homologous parts of all vertebrate brains, so this basic organization is ancient. What most distinguishes the brains of different species of vertebrate animals is therefore not their basic organization, but the relative sizes of the regions. For example, lab rats and ground squirrels have brains of similar sizes, yet the superior colliculus, the part of the dorsal midbrain that is involved in spatial orientation, is ten times bigger in squirrels.[5] The brain cases from fossilized skulls of *Tyrannosaurus rex*, for example, show that they had relatively large brains for dinosaurs of their size, and that they had particularly large olfactory bulbs, for a very keen sense of smell.[6]

Head-to-Tail

We now take a step sideways, away from the brain for a moment, to consider one of the most mind-blowing discoveries in developmental biology. This discovery was made by a diminutive, shy, and wonderful geneticist named Ed Lewis. Lewis worked mostly by himself in his lab at Caltech, mainly at night, and he attracted little attention during most of his long career. The one exception was his 1957 study[7] using medical records from survivors of the atomic bombs dropped on Hiroshima and Nagasaki, Japan, which alerted the world that even the lowest doses of radiation increase the risk of cancer. He was called to testify before a Senate committee about his data, which was then confirmed in other studies. Lewis's early research thus played an important role in stimulating the enactment of various policies regarding radiation exposure. It was, however, Lewis's discovery of a set of genes that provide distinct identities to different segments of the fruit fly embryo that was so extraordinary and won him a Nobel Prize in 1995.

Lewis was a student of Alfred Sturtevant, a colleague of Thomas Hunt Morgan, who worked at the famous Drosophila Genetics laboratory at Columbia University in the early 1900s. The lab collected mutants of the species of fruit fly known as *Drosophila melanogaster* (these are the little guys that are found around your fruit bowl). The Morgan lab had thousands of milk bottles filled with different mutant lines. These mutants were usually discovered by the effects they had on the adult fly's anatomy, such as the presence, size, color, and shape of various body parts. There were wing mutants, eye mutants, leg mutants, bristle mutants, coloration pattern mutants, and so forth. Morgan and colleagues used these mutants to uncover some of the most important principles of genetics. For example, they "rediscovered" Mendel's basic concepts concerning the segregation of dominant and recessive genes based on his experiments with peas in the mid-1800s. Sturtevant was also able to go one step further, which was to locate the mutations to specific regions of particular chromosomes. These regions are where the genes controlling the specific features affected by the mutations are located.

Lewis was particularly interested in mutants in which one body part gets transformed to look like a different body part. This is called "homeosis" (after the Greek for "becoming like"). An example of a homeotic mutant are fruit flies where the antenna appears to be turned into a leg growing out of the front of the head. Lewis studied many homeotic mutations that transformed different segments of the fly into one another. His genetic mapping experiments showed that several of these mutations occurred along a single short stretch on one of the chromosomes. Most astonishingly, the mutations could be arranged in an orderly fashion that linked their positions on the chromosome to their effects on the body. What Lewis found was that the linear order of these mutations on the DNA reflects the rostral-to-caudal (head-to-tail) location of the homeotic transformations they cause in the bodies of the mutant flies. To say this in a slightly different way, neighboring genes affect neighboring segments.[8]

The DNA sequence that contains these mutations was found to have a set of genes whose activity is also organized from head to tail. The first of these genes is active in the most anterior segments, the next is active in the next most posterior segments, and so on. Each of these genes codes for a slightly different version of a transcription factor that shares a sequence of amino acids called a "homeobox" in honor of the homeotic function of these factors. A transcription factor is a protein that binds to specific sites in the cell's DNA and activates or shuts down genes near these binding sites. In this way, transcription factors, like these Hox transcription factors, can influence hundreds of other genes. The fly has eight of these homeobox or "Hox" genes, as we shall now refer to them. The most rostral of the Hox genes works to pattern the head, and the most caudal of the Hox genes patterns the abdominal segments. Lewis noted that loss of any Hox gene causes the transformation of a particular segment in which it was active toward a more mid-thoracic identity. Famously, knocking out the Hox gene that is active in the third thoracic segment transforms that segment into the mid-thoracic segment (where the wings arise), yielding a fly with four wings instead of just two.

The evolutionary ancestors of two-winged flies were four-winged insects, like butterflies and bees. A key to the origin of flies, about 240 million years ago, is the loss of one pair of wings. Mutations in this one Hox gene seem to reverse this ancient evolutionary event. More ancient than insects are the multiseg-mented arthropods like millipedes and centipedes, in which most of the segments are more or less the same. Each of the numerous segments of these creatures has its own pair of legs and its own bit of nerve chord. It is thought that a primordial Hox gene duplicated several times to create a cluster of Hox genes, which became relevant to particular segments in an orderly way. Each Hox gene transforms segments away from their ancient uniform identities into new ones. The Hox genes that are active in the caudal segments of insects eliminate legs and cause these segments to become specialized parts of the abdomen. The Hox genes in the head region transform segments that might otherwise have legs into ones that have typical head structures, like the antennae or proboscis, rather than legs. As do all insects, flour beetles have six legs, but if one deletes all the Hox genes in a flour beetle embryo, the animal that develops is a creature that has 15 pairs of legs, resembling a little centipede more than it does an insect, and one also deletes much of the work of the 400 million years that went into evolving this insect.[9]

Throughout the animal kingdom, during embryonic development, Hox genes are used to differentiate segments or regions of the embryo according to their head-to-tail positions. Like most vertebrates, humans ended up with four Hox gene clusters (A, B, C, and D), each with ten or more Hox genes with names like HoxA1 or HoxB2. The letters A through D indicate which Hox cluster the gene belongs to, and the numbers 1 up to 13 signal the first to the last gene in the cluster. Low-number

Hox genes are active at the anterior regions of the brain, and high-number Hox genes are active at the posterior regions of the spinal cord.

Evolution is not good at creating completely new genes. Instead, evolution takes advantage of preexisting genes and changes them, so they do different things. With four Hox clusters, the genes can repurpose themselves to have more specialized functions, which is why mutations in these genes do not always have the same kinds of effects in vertebrates as they have in flies. Yet the conservation of Hox genes and the fact that they do similar sorts of things in terms of patterning the body is a truly great discovery. For example, when one particular Hox gene is deleted in mice, specific segments of the hindbrain are affected. In humans, mutations in the same Hox gene cause a syndrome called Athabaskan Brainstem Dysgenesis Syndrome, first found in a small number of Native Americans of Athabascan or Dené descent. The human mutation affects the same segments of the hindbrain as it does in mice, leading to deafness, hypoventilation, and facial paralysis, as well as gaze problems that are similar to those seen in mice, as these segments also house many key motor neurons that move the eyes.[10]

Teratogen

The detour we have taken into the world of mutant flies has brought us back to the brain and now leads to the laboratory of the embryologist Pieter Nieuwkoop at the Netherlands Institute for Developmental Biology in Utrecht in the 1950s. Nieuwkoop inserted a flap of as-yet uninduced ectoderm at specific positions along the rostral-to-caudal axis of a host frog embryo at the neural plate or early neural tube stage. He put the flap of tissue into

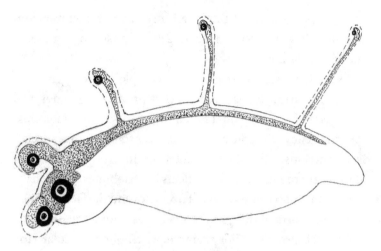

FIGURE 2.3. Nieuwkoop's 1952 experiment. A holistic view of a full set of ectodermal transplants (surrounded by dashed lines) to different head-to-tail positions of the host. All the transplants form neural tissue (stippled) that joins the host central nervous system (also stippled). Each transplant makes forebrain (including eyes) at the tip, but where it joins the host, it assumes the same positional identity as the host neural tube (see text). The same amount of neural tissue is formed from each transplant, so the amount of neural tissue devoted to forebrain becomes smaller with transplants that are closer to the tail.

the neural epithelium of the host, so that its cells would be induced to become neural. The results, which were amazing, can be summarized by two points. First, forebrain structures always developed at the tip, the part of the transplant that was farthest away from the host. Whether the piece was put near the front or the back of the host embryo, it always made some forebrain at the tip. Second, the most caudal (tailward) region that developed in transplanted pieces corresponded to the region of the host embryo to which the piece of tissue was transplanted. For example, when Nieuwkoop put the transplant in the forebrain region of the host, this piece made only forebrain neural tissue (figure 2.3). If the transplant was done at the host midbrain level, the piece developed midbrain structures closest to the host, then forebrain at the tip, and when the transplant was placed into the hindbrain, it made hindbrain there, then midbrain, and finally forebrain again at the tip. To explain this result, Nieuwkoop conjectured that after ectoderm was induced to become neural, the closer the tissue was to the caudal (tail) region of the host, the more it assumed a caudal identity.[11]

Nieuwkoop proposed that there was a gradient of something, highest at the caudal end, which transforms neural tissue to more caudal neural structures. This idea proved to be correct. The first (and still the most potent) molecule discovered to have this activity is known as *retinoic acid*. Embryos of various vertebrate species exposed to very low concentrations of retinoic acid develop more tail and less head, and if they are exposed to slightly higher levels of retinoic acid, they may become entirely headless. Retinoic acid is made from vitamin A by a series of enzymatic reactions inside cells. The highest level of activity in the enzymes that make retinoic acid is found at the caudal or tail end of the embryo. Rostrally (at the head end), cells produce enzymes that break down retinoic acid. This creates source and sink, and therefore a rather stable gradient of retinoic acid, which is high at the caudal end of the neural tube and low at the rostral end. The higher in the gradient a cell sits, the more likely it is to develop a caudal nature.

Now we can tie the effects of retinoic acid to the Hox genes by following the molecules of retinoic acid as they penetrate the membranes of the neuroepithelial cells. Once inside, they find their way into the nucleus, where they bind and activate waiting receptors. These activated receptors turn on Hox genes. Low retinoic acid levels activate only low-numbered Hox genes (i.e., Hox genes that are active in the head), and as the concentration of retinoic acid rises in cells toward the tail, Hox genes with higher numbers get activated. Because higher numbered Hox genes provide more caudal segmental identities, the transforming effect of retinoic acid on the neural tissue is clearly a caudalizing one, as Nieuwkoop predicted.[12] Each Hox gene gets activated at a particular threshold of retinoic acid, so the smooth gradient of retinoic acid becomes divided into discrete segmental identities, each defined by the expression of specific Hox genes.

Vitamin A, found for example in carrots, is necessary for vision. During pregnancy, however, vitamin A deficiency can cause a spectrum of fetal abnormalities due to the loss of retinoic acid. Too much retinoic acid is also bad. Exposure of embryos to high levels of retinoic acid can also interfere with many developmental processes. Levels of retinoic acid have to be carefully controlled. Even a little too much retinoic acid is dangerous for human embryos. This was discovered by the Swiss pharmaceutical company Roche in the 1960s. They found that retinoic acid is an effective treatment for severe acne, but it also produces birth defects if given to pregnant mothers. It is what is known as a *teratogen*, a substance that causes malformations in human embryos. The drug was released, with risk warnings, under the name of Accutane (aka Roaccutane) in the 1980s, but it was still sometimes prescribed to pregnant mothers, resulting in many thousands of babies with severe birth defects, including irreversible brain defects. The use of retinoic acid for the treatment of acne is now much more strictly controlled. For many years, the section of the U.S. National Institutes of Health (NIH) that dealt with teratology supported almost all U.S. research on development neuroscience. The NIH still has no specific institute for developmental neuroscience, though it does have the Institute for Child Health and Human Development, which now supports much of the research in brain development.

Back-to-Belly

The patterning of the neural plate and tube along the rostral-to-caudal (head-to-tail) axis creates segment-like domains of the nervous system. Now consider the lines that run in the other axis on the neural plate, those that divide each of the segments into domains from dorsal (closest to the backside of the embryo) to ventral (closest to the belly). Consider the spinal cord, where different segments control different regions of the body. When you step on something sharp in your bare feet, peripheral pain-registering sensory neurons send signals to the spinal cord to activate motor neurons to lift the foot. The pain-sensitive, sensory axons enter the dorsal part of the lumbar segments of the spinal cord, where they make synapses with local motor neurons. These motor neurons have axons that exit from ventral parts of those segments. A similar withdrawal reflex happens when you touch something too hot with your finger, but this time, the sensory and motor nerves enter and exit the dorsal and ventral parts of the cervical segments of the neural tube. Thus, the segmental rostral (hand) to caudal (foot) patterning of the neural tube is overlain with a dorsal (sensory) to ventral (motor) organization.

The neurons that sense pain, touch, and temperature arise from the dorsal neural tube. Motor neurons arise from the ventral part of the neural tube. It turns out that we are already familiar with morphogens that set up these dorsal-to-ventral (back-to-belly) domains of the neural tube because they are the same molecules that are involved in neural induction. The dorsal part of the neural tube is closest to the overlying epidermis. The epidermis, as you may recall from chapter 1, secretes bone morphogenetic protein (BMP), while the ventral part of the neural tube sits directly on top of the notochord, the primary derivative of Spemann's organizer and the source of anti-BMPs such as Noggin. This creates a dorsal-to-ventral, high-to-low gradient of BMP activity that permeates every segment of the neural tube.

A second morphogen that works as an opposing gradient was discovered through an ingenious and ambitious set of experiments into the way that animals become patterned. Christiane Nüsslein-Volhard, Eric Wieschaus, and their team, working in the Max Planck Institute for Developmental Biology in Tübingen in the 1980s, mutated virtually every single gene in the fruit fly in a search for all the genes that controlled body patterning.[13] They found thousands of mutant lines for embryos that developed to a prehatching stage but were so deformed in some major way that they could not break out of their egg cases. They then manually dissected these tiny mutant embryos from their egg cases and mounted them on microscope slides for detailed examination. What they saw amazed them! There were embryos with heads at either end or tails at either end, embryos where the front half of every segment was turned into the back half, mutants that had missing segments or duplicated segments, and mutants that had various dorsal ventral patterning defects. Hundreds of new genes were associated with such patterning defects. The subsequent study of these genes in various animals transformed the field of developmental biology, which is the reason that Nüsslein-Volhard and Wieschaus shared the 1995 Nobel Prize with Ed Lewis.

Many of the genes discovered by Nüsslein-Volhard and Wieschaus in their search for developmental mutants were named after the defects they cause when mutated, a relatively common practice in the naming of genes. In embryos mutant for one of the new genes, the smooth or naked half of each segment of the larvae was removed, leaving only the hairy or bristle-covered halves. This embryo, when it was removed from its egg case, was short, stubby, and spikey all over, hence the name *hedgehog*. As soon as the fly gene was cloned, developmental biologists working with fish and chick embryos cloned vertebrate versions of this gene and quickly showed that these *hedgehog*-like genes also had roles in patterning the vertebrate embryo. The main vertebrate version of the fly gene is known as *sonic hedgehog* (after the eponymous character of a popular video game). In vertebrate embryos, the sonic hedgehog protein works as a countergradient to BMP (i.e., it is high ventrally and low dorsally in the neural tube). When a piece of the neuroepithelium from the middle of the chick neural tube is put into a petri dish and then exposed to sonic hedgehog, it produces motor neurons; yet if it is exposed to BMP, it makes sensory neurons.

There are many domains from dorsal-to-ventral (back-to-belly) in the neural tube. These dorsal-to-ventral domains arise through the interpretation of the BMP and sonic hedgehog gradients by the cells of the neural tube. An appreciation of how a gradient is broken into discrete domains is afforded by research showing that sonic hedgehog works by regulating pairs of genes in opposite ways. One member of the pair is turned on at a particular threshold level of sonic hedgehog, while the other member of the pair is turned off at the same level. These pairs of genes code for transcription factors that, as well as activating many downstream genes, repress each other. This cross-repression forces cells to activate one or the other of the pair, but never both.[14] As a result, cells belong to one domain or another, and the border between the domains is sharp at a particular threshold of sonic hedgehog. Different pairs of genes respond to different threshold levels of sonic hedgehog, so several such boundaries are drawn. Each domain thereby expresses a unique combination of transcription factors that regulate distinct target genes, which are then used to make particular types of neurons.

Morphogens

Gradients of molecules like retinoic acid, sonic hedgehog, and BMP that work in diffusive gradients to pattern tissue are known as morphogens. Working at University College London, the developmental biologist Lewis Wolpert showed how a gradient of a single morphogen of the sort envisaged by Turing (chapter 1) could be used to pattern a developing animal.[15] Wolpert considered a situation in which one region of the embryo produces a morphogen. This is the "source." The active morphogen then diffuses through the tissue, but a factor that neutralizes the morphogen is released from a different region of the embryo. This is the "sink." This arrangement leads to a high level of morphogen near the source and a low level near the sink. In between the two is a gradient of the morphogen. Wolpert used this concept to explain how such gradients could be used to create the standard proportions, sizes, shapes, orientations, and orders of organ systems in the body. His chosen metaphor was le Tricolore—the blue, white, and red French flag. The source of the morphogen was at the left edge by the blue, and the sink was at the right edge, by the red. As one moves across the flag from the left to the right edge, the concentration of the morphogen decreases. Now, imagine that the flag is full of cells that can sense the concentration of the morphogen. When they sense a concentration above a particular threshold, they activate the "blue" gene; below this threshold but above a lower threshold, they activate the "white" gene, while the "red" gene is activated in the default state, in which the level of morphogen is below the threshold for activating either the blue or white genes. As long as there is an effective source and sink at the two edges and an even gradient in between, the flag of cells would retain its blue-white-red proportions whether it was large or small.

As the embryo grows and the neural tube becomes larger and longer, new sources and sinks of morphogens arise to pattern the nervous system. For example, sonic hedgehog is first made by the notochord, which lies under the ventral midline of the neural tube. Consequently, this region of the neural tube is exposed to the highest level of sonic hedgehog. Similarly, the most dorsal part of the neural tube is exposed to the highest level of BMP, which is made by the overlying epidermis. As a result of these exposures, the most dorsal part of the neural tube itself begins to express BMP, and the most ventral part of the neural tube begins to express sonic hedgehog. With growth of the embryo, the original sources of these morphogens (notochord and epidermis) become increasingly distant from the neural tube, and the new local signaling centers become more crucial.

The junction of the midbrain and the hindbrain is another key local signaling center in the developing neural tube. This junction, which is at the site of one of the first constrictions in the neural tube, makes morphogens with important organizing influences on the formation of adjacent brain regions. If small pieces of this border region are transplanted from one chick embryo into the forebrain region of another, an extra cerebellum arises from the host cells on one side of the transplant, and an extra midbrain forms from host cells on the other side of the transplant. In other words, this small piece of neural tissue is able to organize the brain regions around it, probably by releasing one or more morphogens. One of the key morphogens released by the cells of the midbrain/hindbrain border region is a small, secreted protein known as Wnt (pronounced "wint"). Wnt was originally discovered through research into why some viruses (called "tumor viruses") cause cancer. In 1983, Roel Nusse and Harold Varmus were investigating a mouse mammary tumor virus. They found that when the virus infected cells, it made a DNA copy of itself, and this copy often "jumped" into the host cell's own DNA. Nusse and Varmus searched for the regions of DNA where this integration occurred and led to tumors, and they called the first such site of integration that they found "int-1."[16] They speculated that the integration of viral DNA at the int-1 site enhanced the expression of a nearby cancer-causing gene. They were correct. The gene near int-1 turned out to be a homologue of a secreted protein that had already been identified as responsible for a mutant line of *Drosophila* that had no wings. The gene in flies was called (you guessed it) *wingless*, and so the int-1 gene was named Wnt1. Humans have about 20 distinct Wnt genes. When the Wnt1 gene is deleted in mice, the animals cannot lose their wings, of course, but they do lose most of their midbrain and cerebellum. The Wnt released at the midbrain/hindbrain junction is opposed by anti-Wnt molecules, some colorfully named after the defects seen when they are not functioning properly, such as "Cerberus," after the three-headed dog of Greek mythology that bars the living from entering the gates of the underworld; and "Dickkopf," the German word for "thick-headed" or "stubborn." The production and release of Wnt at the midbrain/hindbrain border, and the production and release of anti-Wnts at the front or rostral end of the neural tube, creates a source and sink for this powerful morphogen and so creates a local morphogenetic gradient that helps pattern the front of the brain.

The result of all this head-to-tail, dorsal-to-ventral, and new local sources and sinks of morphogens is the division of the neural tube into groups of neural stem cells with specific regional identities. The remarkable similarity between flies and mice, in the way that head-to-tail and dorsal-to-ventral morphogens regulate certain transcription factors at distinct thresholds of these morphogens, suggests that these are evolutionarily ancient attributes of brain patterning. Perhaps insects and vertebrates inherited their basic Bauplan from a common ancestor that used similar developmental mechanisms.

To gain insight into this idea, Detlef Arendt at the European Molecular Biology Laboratories in Heidelberg has been studying the nervous system of a marine worm called *Platynereis*. These polychaete worms are segmented animals in which each segment has a pair of bristly appendages. The fossil record places the origin of this group of worms near the beginning of the Cambrian Period, approximately 515 million years ago, from a branch of the evolutionary tree prior to the origin of vertebrates. *Platynereis* retains a strong morphological likeness to these early fossils, suggesting that modern representatives retain many attributes of their ancient ancestors. The nervous system of *Platynereis* worms is arranged as a segmented nerve cord running the length of the body, with a brain and sensory organs, such as eyes, in the head. Arendt and colleagues showed that *Platynereis* embryos have counterparts of many of the same patterning genes that were first found in *Drosophila* by Nüsslein-Volhard and Wieschaus, and these genes work to pattern the nervous system.[17] It is a sobering thought that an ancient program of nervous system patterning leads to the development of different regions of the brain, with specific conserved functions at similar relative positions in the brains of animals as diverse as flies and humans.

Many of the developmental pathways that are discussed in this and other chapters are ancient and have been used again and again for similar purposes in different animals. But it is worth reiterating that, as is the case for the Hox genes, many of the morphogens that are used in patterning the brain have also been repurposed to do new things as the brain has evolved. For example, we learned in chapter 1 that during neural induction, BMP causes cells of the ectoderm to become epidermal rather than neural, but in this chapter, we learned that BMP is rapidly repurposed to pattern the dorsal neural tube without causing it to become epidermis. Later it is used again to help subdivide the other parts of the brain. Wnt is used in patterning the brain into different regions, but it is also used to drive the growth of tissue. Indeed, this is why it was first discovered as a cancer gene. One gets the impression that building different organisms is like playing Rachmaninoff, Beethoven, or a boogie-woogie on a piano, using the same keys in different ways and combinations. One does not need new genes to make new structures.

Eyes

Let us start looking for the origin of a very specific part of the nervous system, the retinas of our two eyes, by fearlessly considering the embryonic life of the one-eyed, giant cyclops called Polyphemus, whom Odysseus lured into a drunken stupor before shoving a sharpened stake into his one central eye. Polyphemus was one of several mythical cyclops who herded sheep in Sicily. In 1957, in the very real world of Idaho, sheep farmers found a high incidence of cyclopic lambs, born with one eye in the middle of their heads. With the help of the Department of Agriculture, the culprit was found to be a chemical that is naturally produced by California corn lily (*Veratrum californicum*), on which the pregnant sheep were sometimes grazing. This chemical substance was called "cyclopamine" for obvious reasons, and its mechanism of action was unknown for many decades. However, once sonic hedgehog had been discovered and its morphogenetic effects known, it became possible to discover that cyclopamine's main action is the inhibition of sonic hedgehog signaling.[18]

In the embryos of the pregnant sheep, as in all vertebrates, the two eyes originate from a single domain, called the "eye field," which sits in the front and middle of the neural plate. As the neural plate rolls into a tube, two ventral bulges begin to appear in the forebrain region. These bulges, the left and right wings of this single eye field, are becoming the buds of the left and the right eyes. The division of the single eye field into these two bulges depends on the sonic hedgehog being released at the midline. In Polyphemus, when he was an embryo, perhaps because of a mutation in the sonic hedgehog signaling pathway or perhaps because his mother's diet incorporated cyclopamine, the eye field did not split, and the result was cyclopia. In humans, almost all such cases are incompatible with life.

The beautiful eyes of a baby start out in this central single eye field, which has been specified by transcription factors that determine the intrinsic character of the eye. If there is a master of eye development, the gene called Pax6 is certainly a contender. If both copies of the Pax6 gene are mutant in a mouse embryo, it is born with no eyes. The Pax6 gene is turned on in the eye field of all vertebrates so far tested. The protein is highly conserved among mammals. The mouse version, for example, is identical to the human version. Exactly how well conserved Pax6 is among all animals came as a shock to the developmental biology community when it was discovered by Walter Gehring's group in 1994 that a fruit fly mutant called "eyeless," known since the 1920s because (you guessed it) it has no eyes, has a mutation in the fly's version of the Pax6 gene. The high level of conservation of Pax6 gene function was stunningly demonstrated by gene transfer experiments.[19] These showed that the human Pax6 gene can substitute for the fly's own version of the gene and rescue the otherwise eyeless mutant fruit flies. This was particularly surprising because until this point, most biologists assumed that the compound eyes of insects evolved completely independently from the camera-like eyes of vertebrates. Indeed, it was thought, given their great variety, that eyes may have evolved independently up to 40 times in the animal kingdom. Now, though, it is thought that simple eyes may have evolved only once and then diversified into a myriad of forms.

It takes just the right set of transcription factors to turn on all the genes needed to transform tissue into eyes. Flies have many other genes, which when mutant, also lead to the failure of eyes to develop. Not only "eyeless," but also genes with names like "twin of eyeless," "eye-gone," "eyes absent," and "sine oculis" (Latin for "without eyes"). These eye genes code for transcription factors, and many of them activate each other, thus forming a self-regulating molecular pathway for turning on all the genes that make the eye. Because the network is self-regulating, the activity of almost any one of them can kick-start the whole network into action. One of the most dramatic demonstrations of this came when Gehring's lab intentionally directed the Pax6 gene to be active in such regions of the developing fly as the antennae, legs, wings, and genitalia. The result was a fly in which all these appendages were transformed into eyes. Imagine a fly with 15 eyes all over its body, but no legs, wings, antennae, or genitalia! Similarly, if the frog versions of these eye-field genes are experimentally activated in different parts of a developing frog embryo, it can cause extra eyes to develop even in the belly or tail region![20]

The eye field of the vertebrate neural plate, like other areas of the brain, arises because it is in exactly the correct position within morphogenetic gradients that pattern the neural tube to turn on all the necessary transcription factors, like Pax6. Knowledge of the morphogenetic signals and the master regulators that are involved in making eyes is allowing modern scientists and clinicians to produce retinal tissue from human embryonic stem cells in tissue culture dishes. In this way, we can generate cell types that are useful for repairing retinas. Using stem cells derived from patients, researchers can now also make retinal cells, such as rods and cones, that are genetically identical to the patient's own cells for medical testing or potential replacement.

Areas of the Cerebral Cortex

In 1909, the neurologist Korbinian Brodmann, working at the Neurobiological Laboratory at the University of Berlin, identified dozens of distinct areas of the cerebral cortex in a variety of mammals.[21] He did this by looking at sections of the cortex under a microscope and noting their histological characteristics (e.g., the arrangements and numbers of different types of neurons in the different layers of the cortex). Brodmann found 52 histologically distinct areas of the human cerebral cortex. Many of these areas, now known as Brodmann areas, correspond to regions that process specific types of information (figure 2.4). For example, Brodmann area 1 is somatosensory cortex, area 17 is visual cortex, and area 22 is auditory cortex. Although Brodmann's scheme has held up remarkably well over the years, we now know there are many more distinct functional areas in the cortex than even Brodmann saw (see chapter 9). Each area is endowed with its own neural architecture, its own organization, its own way of functioning, and its own inputs and outputs

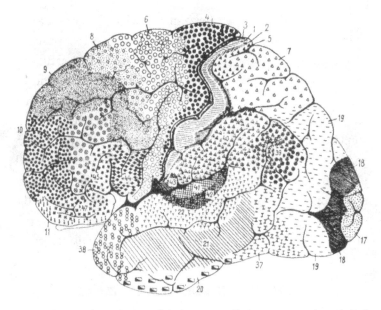

FIGURE 2.4. Brodmann's areas of the human cerebral cortex. Lateral view of left cerebral cortex.

from and to other brain regions. The borders between some of these regions are sometimes gradual and sometimes sharp, and many of these areas and their boundaries can be seen in brains of fetal humans in the second trimester.

Macaque monkeys have many of the same cortical regions as humans, and these regions are arranged in a similar way across the cortex, though their cerebral cortex is ten times smaller than ours, and even mice have many of the same cortical regions arranged with the same general relative topography. This suggests that these cortical regions develop through shared mechanisms (such as morphogenetic gradients) that establish borders of gene activity, which in turn drive the distinguishing characteristics of one cortical area from another. Work in mice strongly implicates the existence of several local patterning or signaling centers in the developing cerebral cortex. For example, one of the morphogens, called "fibroblast growth factor" (FGF), is normally produced at the front edge of the developing cerebral cortex. If extra FGF is experimentally given to the front part of a mouse brain during early cortical development, the cortical areas at the front of the brain, such as the motor cortex, become larger, displacing areas like the somatosensory area (which is normally near the middle of the cortex) rearward, and compressing areas at the back of the brain, like the visual cortex. If a source of FGF is experimentally positioned at the back of the brain, so that the animal has two gradients, one coming from the front and one from the rear, many areas of the cortex become duplicated. In this case, there are two somatosensory and motor areas, while the visual cortex, usually at the back of the cortex but now banished from both the front and the back, is displaced to the middle.

If borders between areas and domains were drawn at one specific time, such variations in the shape of the gradient at that time would likely lead to larger variabilities in a region's size and position than is normally seen between regions. And although cells use a variety of mechanisms to reduce this variability, such as integrating signals over time rather than making decisions based on an instantaneous snapshot of the gradient, exactly where the boundaries are drawn will almost certainly be slightly different between individuals. The exact shapes and sizes of cortical areas are indeed quite variable and can be used to distinguish human beings from one another (chapter 8).

Variation is a key element of evolutionary theory, and it is useful to consider the cerebral cortex in this context. It is known that the relative sizes of regions of the cerebral cortex are different for different mammals. The cerebral cortex of the hedgehog (the mammal, not the morphogen!), is one of the most primitive among mammals. Hedgehogs have poor vision, and a relatively small amount of its cerebral cortex is devoted to vision, with only a few cortical areas for interpreting visual signals. In contrast, primates like us and macaque monkeys are highly visual mammals, and our cerebral cortex has more than 30 different visual areas. These regions are differentially tuned to different aspects of the scene: movement, color, distance, context, and so forth. Other mammals specialize in other senses. For example, bats navigate and hunt using echolocation: hearing the reflections of their high-pitched cries. They have a relatively large cortical area devoted to hearing, with several cortical areas specifically concerned with processing different aspects of echoes. Raccoons explore the world with their hands, and mice and rats use their whiskers, and the amount of cortex devoted to these somatosensory inputs is relatively enlarged in these animals. These differences in the cortical areas in different animals are not due to postnatal sensory experience but to evolution and development. The mechanisms behind these changes are still relatively unexplored, but it makes sense that the morphogenetic gradients and transcription factors discussed throughout this chapter participate in drawing the borders between these areas of the cerebral cortex.

By one month of gestation, the human neural plate has rolled itself into a tube. This neural tube has become patterned along the head-to-tail axis and back-to-belly axes, such that the embryonic origins of different parts of the adult nervous system have been specified by mapping regional coordinates on the neural tube. The patterning mechanisms involve gradients of diffusible signals called "morphogens" that act at specific thresholds to activate or repress genes that code for transcription factors. The smooth gradients are transformed into sharp boundaries that separate different regions of the brain. As the brain grows, new local signaling centers arise and create new gradients that subdivide the brain further into a patchwork of distinct regions. This conceptual framework holds true for vertebrates and invertebrates, as it evolved in our common ancestors. Indeed, much of the molecular logic for generating specific parts of the nervous system has been highly conserved over hundreds of millions of years of evolution. Variability in the morphogen gradients as well as the retuning of the transcription factors that they regulate hold clues to the evolution of brain specializations among different species. While the brain is being divided and subdivided into its many regions, it is also growing at a tremendous rate. Soon there will be tens of billions of neurons. Chapter 3 explores how the neural stem cells become such a multitude.

3

Proliferation

In which we learn how the growing population of
neural stem cells produces a brain of the right size
and proportions, and we consider whether there are
any neural stem cells in the adult human brain.

Proliferation

During the first four weeks of gestation, the human embryo has
a population of thousands of neural stem cells. The neural
tube is packed with them, and they are rapidly proliferating. A
typical newborn baby's brain has about 100 billion neurons,
so a lot of proliferation is needed. As birth approaches, growth
slows down because more and more of these cells turn into
neurons—which do not divide. By birth, neuron production is
complete in almost every region of the brain. How does this
happen? How is the growth of the brain and all its regions controlled so precisely? And what can go wrong?

When they first arise, neural stem cells divide to produce
two more neural stem cells with each division. This symmetrical proliferative type of division leads to an exponential increase in neural stem cell numbers. It is estimated that near the

end of the first trimester of pregnancy, the fetal human brain is producing 15 million cells per hour. In the second trimester, many neural stem cells change their mode of division. In this new asymmetric mode, when a neural stem cell divides, it produces one daughter cell that remains a neural stem cell like its mother, but the other daughter is different. It either turns into a neuron immediately or becomes a secondary progenitor that divides a very limited number of times before producing a small cluster of neurons. As more neural stem cells switch to this asymmetrical mode of division, the increase of cell numbers becomes less exponential and more linear. By the third trimester, the neural stem cells change their mode of division once more. They now often divide to produce two daughters that are both neurons, neither of which will ever divide again, and neuron production begins to flatten out. Once the last of these neural stem cells has finished dividing, neural proliferation is complete. By the time of birth, the increase in neuronal numbers has virtually ground to a halt.

Even though the head of every newborn baby contains a brain with almost all the neurons it will ever have, the brains of newborns are only about one-third the size of adult human brains. It takes another six or seven years before human brains reach their almost-adult size. If all the neurons exist by the time of birth, what accounts for the tripling in the size of the brain during childhood? Some of this increase is because the young neurons themselves get larger after birth; they continue to grow as they send out branches and make new connections. But perhaps the biggest factor in postnatal brain growth is that the non-neuronal cells of the brain, known as glial cells, continue to be generated postnatally. One type of glial cell, the oligodendrocyte, makes the white matter in the brain. Oligodendrocytes wrap the long axons of neurons with multiple layers of lipid-rich membranes, which look whitish when packed densely. This wrapping, called "myelin," serves to insulate axons against leakage of electrical current, allowing nerve impulses to travel farther and faster. White matter accounts for about 40% of the mass of the adult brain and is mostly generated postnatally. Another major type of glial cell in the adult brain is the star-shaped astrocyte. Next to neurons, astrocytes are the most numerous cell type in the brain. Astrocytes do many jobs. They provide a conduit between the blood supply and the neurons that use its nutrients and oxygen, they maintain an optimal ionic environment for electrical signaling, they participate in the formation and maintenance of synaptic connections between neurons (see chapter 6), and they support brain function in general.

The number of neurons in a brain is a product of how many rounds of cell division neural stem cells go through. More rounds of division produce more neurons, but it takes a longer time for such lineages to play out. It takes about four months for a human fetus to generate the 25 billion neurons of the human cerebral cortex. In comparison, a macaque monkey fetus generates the 1.5 billion neurons of its cerebral cortex in two months, and the tiny cerebral cortex of a baby mouse with only about 15 million neurons is generated in just ten days. Because neural stem cells are packed side by side, each one being attached to both the inner and outer surface of the neuroepithelium, the surface area of the neuroepithelium doubles with each cycle of division. So it only takes three extra early rounds of cell division to expand the surface area of the cerebral cortex by a factor of eight. Once these cells have finished their symmetric proliferative cell divisions, they go through rounds of asymmetric cell divisions that contribute to the growing thickness of the cortex. Of the cortical neural stem cells in these three species, the human ones go through the most rounds of symmetrical and the most rounds of asymmetrical cell divisions. A factor that distinguishes primate neural stem cells from those of the mouse relates to the "other" daughters of asymmetrical divisions, the secondary progenitors. In mice, these other daughters either become neurons directly or divide only once more, generating two neurons. In humans and monkeys, the secondary progenitors may divide several more times before their progeny turn into neurons.[1]

An inquisitive person might ask if a neural stem cell from a mouse were put into the developing human cortex and thereby exposed to the nutrients and other factors that are available there, would it proliferate more like a human cortical stem cell? Rick Livesey did an experiment like this at the University of Cambridge in 2016. Instead of gathering the cells from embryonic animals, Livesey made cortical neural stem cells from totipotent cell lines (see chapter 1). Livesey and colleagues then compared the lineages of stem cells of the cerebral cortex generated in this way from humans, from marmoset monkeys, and from mice when they were put into the same culture conditions.[2] The result was clear: the human cortical stem cells remained proliferative for a longer period, went through more rounds of division, and so made the biggest clones of neurons. Monkey cortical stem cells divided fewer times and made smaller clones, while mouse cortical stem cells divided the fewest times and made the smallest clones. Even when the monkey stem cells were mixed in among the human stem cells, the monkey stem cells made monkey-sized clones, and human stem cells still made human-sized clones. This result suggests that these cortical stem cells are intrinsically programmed to proliferate in a species-specific manner. The stem cells of the cerebral cortex cells seem to "know" that they are mouse, marmoset, or human.

These studies complement those done by the Stanford experimental embryologist Victor Twitty in the late 1930s. Twitty captured my scientific imagination with his wonderful little book *Of Scientists and Salamanders*.[3] In one set of experiments, Twitty describes how he exchanged a limb bud from an embryo of a species of large salamander with a limb bud from an embryo of a smaller species. The embryos with transplanted limb buds grew into adults that were amazing to behold. Here was a small salamander sporting a leg almost as big as the rest of its body, and there was a big-adult salamander with three normal legs and a fourth tiny one. The same type of results pertained to the retina, the part of the brain that sits in the eye. When Twitty exchanged eye primordia between the embryos of big-eyed and small-eyed species, the eye primordia from the big-eyed species still grew into big eyes with large retinas and lots more neurons, even though Twitty had transplanted them into the embryos of the small-eyed species, and vice versa. These results point to an intrinsic, species-specific component that governs the proliferative potential of neural stem cells.

Invariant and Variable Lineages

To know a cell, you should know its lineage. Who was its mother, who was its grandmother, and so on? In the late 1970s, the tiny soil nematode of the species *C. elegans* was chosen by John Sulston and colleagues for an enormous scientific challenge. They wanted to know the complete lineage history of every cell in the body. They chose this species because it is composed of less than 1,000 cells, and one can see them all under the microscope in a living embryo. Through truly heroic efforts and countless hours of tracking every cell in time and space, they managed to follow every cell division from egg to adult and thus trace the entire ancestry of each cell all the way back to the egg.[4] They then compared the results from one animal to the next, and by doing so they discovered that every *C. elegans* nematode reenacted the same lineage histories. It seems that every cell division in every lineage plays out in the same way in every nematode. Such invariant cell lineages are found often in invertebrates, and especially among those that have few neurons, such as this tiny nematode, whose nervous system is composed of exactly 302 neurons and 56 glial cells.

It is much more challenging to trace the lineage histories of neurons in the brains of larger vertebrate animals, which have orders of magnitude more neurons. It has, however, proved possible to trace parts of their lineages. This has been enough to show that vertebrate neural lineages are much more variable than the lineages seen in nematodes. For example, one stem cell in the cerebral cortex of a mouse embryo might generate a clone of hundreds of neurons, and its seemingly equivalent neighbor might generate less than 20. How is it possible to make brains of the right size and proportions if neural stem cells show such variability in the numbers of neurons they produce?

One possibility is that a certain amount of randomness occurs in the way that neural stem cells of vertebrate brains behave. One illustration of how this works comes from the time-lapse movies that Jie He made of retinal stem cells in transparent zebrafish embryos when he was in my lab at Cambridge University.[5] With a microscope, one can look into the living brain of a zebrafish embryo as it is developing and follow single cells through many divisions. There are about a thousand retinal stem cells in a zebrafish embryo. In my lab, Jie He traced the lineages of hundreds of retinal stem cells. Most retinal stem cells in these fish start with three rounds of exponential growth, in which each retinal stem cell divides into two more retinal stem cells. Each of these eight great-granddaughters then enter a fourth round of cell division, in which it seems to choose randomly between one of three different division modes. In the first mode, the cell divides as it had before (i.e., symmetrically to make two more proliferative cells). In the second mode, the cell divides asymmetrically to make one proliferative cell and one cell that becomes a retinal neuron, which never divides again. In the third mode, the cell divides to generate two retinal neurons. After this fourth round of division, any remaining proliferative cells are strongly biased to make only one last division. After this, retinal proliferation is finished, and the embryonic retina has reached its final size. Because there is a random choice in the three modes of cell division during the fourth round, clone sizes are variable. Yet even though some zebrafish retinal stem cells make small clones and others large clones, the total number of retinal neurons produced is remarkably consistent from fish to fish. The law of large numbers says that in situations where there is a random element at play (like rolling dice or choosing one of three modes of cell division), the average result of performing the same experiment a large number of times will be close to the expected value. As there are about 1,000 retinal stem cells in the zebrafish retina, it is like doing the same experiment on a zebrafish retinal stem cell 1,000 times. The law of large numbers predicts that the total number of cells in the zebrafish at the end of the embryonic period should be close to 24 (the average clone size) times 1,000 (the number of retinal stem cells), that is, 24,000 cells, which is approximately right. Because the "experiment" is done about 1,000 times in each retina, the variation in size between zebrafish retinas is small. The variability in clone sizes for the retina and the cerebral cortex in a variety of mammalian species (including humans) points to a similar reliance on the law of large numbers to build a brain of the right size and proportions out of the somewhat unpredictable patterns of proliferation among individual neural stem cells.

The Cell Cycle

A cell division marks a moment in the life of a neural stem cell, or any dividing cell, when it splits in two. But as each division halves the size of the stem cell, to divide again, it must grow between divisions. The stem cell dance is simple: grow, then divide, then grow again, then divide again, and so on (figure 3.1). This dance is called the "cell cycle." The cell cycle has four phases.[6] The first phase is all about growing and getting enough supplies to be able to commit to the second phase of the cell cycle. Having successfully completed the first phase, the cell enters the second phase, during which it replicates all its DNA. During the third phase, the cell checks for errors in DNA replication and makes repairs to any DNA damage. Once these checks and repairs and some further growth have been completed, the cell moves to the fourth and final phase, called "mitosis," where the two copies of the DNA separate toward opposite sides of the mother cell. The cell then constricts in the middle until the two halves are completely separated from each other. There are now two distinct daughter cells that immediately enter the first growth phase. The mother cell has ceased to exist—it is an ex-cell.

Between the phases of the cell cycle, there are checkpoints. These checkpoints are temporary pauses for quality control, that is, times when the cell checks to make sure one phase is complete and that it is fully prepared for the next phase. Progression through cell-cycle checkpoints is policed by proteins that either promote or inhibit entry into the next phase. If

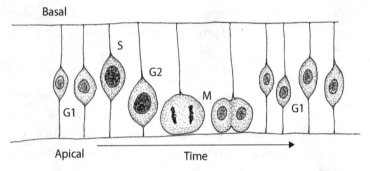

FIGURE 3.1. Cell cycle of a neural stem cell. The dividing cell is part of the neuroepithelium; as such, it retains connections to both the apical and basal surfaces of the neuroepithelium throughout. From left to right: in the first growth phase (G1), the cell grows. Once it has grown enough, it enters the synthesis phase (S), where it replicates its DNA. It then moves to the second growth phase (G2), where it grows a bit more and moves toward the apical surface to divide during mitosis (M). The two daughter cells then reenter the cell cycle in G1.

checkpoint policing proteins do not do their jobs properly, cells can divide when they should not divide. The checkpoint between the first and second phase is a particularly well-studied one because this checkpoint is where most cells get stopped. One protein that polices this major checkpoint is the product of the "retinoblastoma" gene. Without adequate retinoblastoma protein, cells move through the checkpoint too easily. Retinoblastoma is named after a cancer found in babies and young children in which neural progenitors or immature neurons in the retina remain in the cell cycle when they should already have ceased division. This neural cancer occurs in the very young because once neurons have fully differentiated, they do not reenter the cell cycle. Thus, if the tumor is removed, the child usually survives, though having a defective retinoblastoma gene increases the risk of cancers in other tissues where the cell cycle remains active.

In 1931, the physiologist Otto Warburg won the Nobel Prize for his work on tumor metabolism. The problem that puzzled Warburg was that solid tumors begin to grow without their own blood supply. Tumors eventually do attract blood vessels, which fuel their further dangerous growth, but at the beginning, without a blood supply, the cells of a tumor have few nutrients, and yet they manage to proliferate. Warburg wanted to know how they manage to do so. What he found was that cancer cells burn fuel in a different way from that of most cells in the body.[7] Most of our cells metabolize glucose using oxygen, breaking it down to carbon dioxide and water in a way that produces energy stores for the cell. However, during sprinting or powerlifting, muscles burn fuel without oxygen (i.e., anaerobically) because the blood supply is insufficient to deliver all the oxygen that is needed to burn the fuel completely during such sudden bursts of work. Anaerobic metabolism is only about 4% as efficient as aerobic metabolism in terms of producing energy stores, and it also has the distinct disadvantage in muscle cells of building up intermediate products of glucose metabolism, such as lactic acid, which cause muscle pain. Warburg noted that cancer cells also burn fuel anaerobically, but they do so even when there is plenty of oxygen available. Out of the partially burned fuel, including lactate, the cancer cells can make new amino acids and nucleotides and so grow more efficiently than they would if they burned the fuel all the way down to carbon dioxide and water. "Warburg metabolism" not only helps cancer cells grow, but it also frees them from the need for a good supply of oxygen.

A typical human fetus develops in an environment with very little oxygen. The amount of oxygen in the blood of a fetus is comparable, in fact, to that of human breathing without supplementary oxygen on the summit of Mount Everest. The developing placenta and specialized molecular transport machinery create ingenious links to deliver some oxygen from the mother's blood to the fetus, but the levels of oxygen in the embryo are still incredibly low, so perhaps it is not surprising to learn that the highly proliferating stem cells in embryonic tissues often use Warburg metabolism to help them grow enough to safely pass through their cell-cycle checkpoints.

Although embryos can grow in relatively low levels of oxygen, they cannot grow when nutrient levels are too low. Without nutrients, cells cannot get through the first checkpoint, which is why maternal undernutrition leads to undersized babies. Surprisingly, undernourished babies, while small overall, appear to have relatively large heads for their body size. This is due to a phenomenon known as "brain sparing." Perhaps, because the brain is such a crucial organ, and perhaps because neurons, unlike most cell types, are not replenished during a lifetime, starving embryos mobilize their limited resources for building the brain as a top priority. In these cases, the fetal brain grows at the expense of the rest of the body. Nevertheless, children of malnourished mothers often also have smaller-than-average brains, with fewer neurons and glia, and they may show irreparable cognitive disabilities.

The neural stem cells of different regions of the brain proliferate to different extents, so that different regions grow to different sizes with their own populations of neurons. The boundaries between regions of the brain are drawn in the neuroepithelium by thresholds of morphogens and regional transcription factors (see chapter 2). These factors that pattern the brain can also control the differential proliferation of regions of the brain. They exert this control by regulating components of the cell-cycle machinery, such as those that operate as policing factors for checkpoints to ensure that these different regions grow to their appropriate sizes. For example, for the retina to grow to the correct size, at least some of the transcription factors that are key to the formation of the eye (chapter 2) inhibit factors that guard cell-cycle checkpoints, driving the growth of the eye. If the activity of these transcription factors is lost or diminished in frog or fish embryos, the result is that such animals grow tiny eyes, but if these same transcription factors are overactive, the result is formation of giant eyes.[8]

Microcephaly

Schlitzie Surteez, formerly known as "Schlitzie the pinhead," was a side-show entertainer and later a movie star. He died in 1971 at the age of 70. Schlitzie had microcephaly (from the Greek for "small head"). Microcephaly results from a disorder of neural stem cell proliferation. Schlitzie was small, only 122 cm tall, probably because cell proliferation was affected in all the cells in his body to some extent. However, as in most cases of microcephaly, Schlitzie's brain was disproportionately reduced compared to the rest of his body. Though intellectually impaired, he is said to have retained a wondrous and child-like view of the world, akin to that of a three-year-old. He was sociable; and he loved to sing, dance, and entertain, which is why his fellow performers and caregivers protected him as best they could throughout his long life.

Hereditary microcephaly runs in families throughout the world, which has made it possible for geneticists to hunt for and find causative genes. Approximately 20 genes have so far been found that are associated with hereditary microcephaly. Surprisingly, most of these genes code for protein components of a single cellular organelle, called the "centrosome."[9] The centrosome is involved in the final phase of each cell cycle (i.e., mitosis). At the beginning of mitosis, the centrosome splits into two parts, which move to opposite sides of the mother cell. There they anchor themselves and begin assembling a spindle-shaped structure built of protein cables known as microtubules. With the aid of this spindle, motor proteins in the cells begin the physical work of separating the two sets of replicated chromosomes from each other and pulling them along the spindle to each side of the mother cell. The cell division that follows is perpendicular to the orientation of the spindle.

Recall that in the symmetric proliferative mode, the number of cells grows exponentially; in the asymmetric mode, it grows linearly. The probability and time of switching from symmetric to asymmetric therefore has a major impact on brain growth and the final size of the brain. Every neural stem cell contains some factors that bias the cell toward exiting the cell cycle, as well as other factors that favor remaining in the cell cycle. It is as if two opposing forces were wrestling with the cell's soul. One cries, "Stay young! Keep dividing!" The other says, "Grow up! Become a neuron!" If these two types of opposing factors are divided unequally between the two daughters, one daughter may stay in the cycle, while the other differentiates into a neuron. If, however, these factors are split equally between the two daughters, the subsequent division is likely to be symmetrical again.

All neural stem cells have a polarity. They are aligned with the inside-to-outside axis of the neuroepithelium. The part of the cell that is closest to the inside surface is called "apical," and the part of the cell closest to the outside surface is called "basal." Factors that favor remaining in the cell cycle and other factors that favor leaving the cell cycle are often found close to the apical and the basal poles of the cell (figure 3.2). If the cell divides at an angle that is well aligned with its apical-to-basal polarity, the division will split the factors equally between the two daughters, and the division will likely be symmetric. If, however,

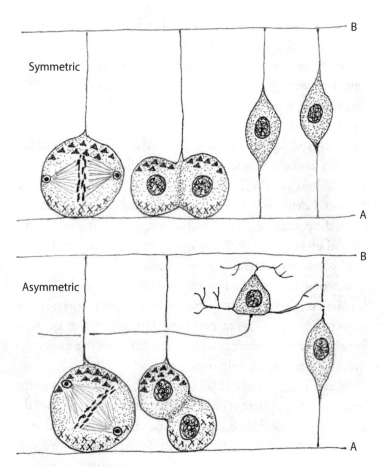

FIGURE 3.2. Symmetric and asymmetric cell divisions. In the top panel, a neural stem cell divides symmetrically, with the mitotic spindle parallel oriented to the apical (A) surface. The two daughter cells divide key molecules (triangles and Xs) evenly between them. Some of these molecules tend to keep the cell in the cell cycle, while others favor exiting the cell cycle. In this symmetric division, both daughter cells reenter the cell cycle. In the bottom panel, the spindle is oriented at an angle to the apical (A) and basal (B) surfaces. As a result, the key molecules are divided unevenly between the two daughter cells, and one daughter becomes a neuron, while the other remains a neural stem cell.

the angle is misaligned, as it often is when the centrosome is compromised, the factors are split unevenly, resulting in asymmetric rather than symmetric divisions and thus limiting the extent of neural proliferation, which is thought to be one of the main causes of hereditary microcephaly.

Several of the genes implicated in hereditary microcephaly have evolved rapidly in bigger-brained mammals (e.g., chimpanzees, humans, whales), implicating centrosome function as a possible key to the big size of our brains. It is not all genetic, though. Environment has a role to play. Microcephaly can be caused by alcohol abuse during pregnancy. It can also be caused by infection with the Zika virus, as found during the 2016 pandemic, when it was first noticed in Brazil that pregnant women who became infected with Zika virus had a high incidence of microcephalic babies. Subsequent research has shown that Zika virus targets neural progenitors in the fetal brain and impairs their proliferation, although the molecular mechanisms are still unknown.

Microcephaly is due to abnormally reduced growth of the brain, while "megalencephaly" is due to excessive growth of the brain. Megalencephaly is caused by mutations that lead to over-activation of cell proliferation, causing neural stem cells to divide more than they should. Note that megalencephaly is not the same as "macrocephaly," which usually stems from a problem with the circulation of the cerebrospinal fluid, leading to the buildup of fluid in brain ventricles or in the space between the brain and the skull, not the over-proliferation of neurons. As this extra fluid can be drained or shunted, babies born with macrocephaly tend not to have long-term neurological problems, whereas a large fraction of babies with definitive megalencephaly at birth are later diagnosed with autistic spectrum disorder. There are no cures yet for either microcephaly or megalencephaly.

The Layers of the Cerebral Cortex

At some point in the development of the brain, the dividing neural stem cells become nondividing neurons. The birthdate of a neuron corresponds to the moment that it leaves the cell cycle forever, never to divide again. Most neurons are born on the inner or apical surface of the neuroepithelium (also called the "ventricular surface," as it lines the ventricles of the brain). The outer or basal side of the neural tube is called the "pial surface," as this is where the first covering of the brain, known as the pia mater (Latin for "tender mother"), wraps the brain in a delicate, translucent sheet of membrane. In the cerebral cortex, the first-born neurons detach themselves from the inner (ventricular) surface of the neuroepithelium and begin to migrate toward the outer (pial) surface. As more cells leave the cell cycle and make this migration, the neuroepithelium becomes separated into an inner zone next to the ventricular surface (where the neural stem cells are still proliferating) and an outer zone near the pial surface (which is filling up with neurons). Microscopic images and time-lapse movies of the developing cerebral cortex show that when a neuron is born in the ventricular zone, it grabs onto one of its neighbors, an elongated neural stem cell that is attached to both the inner and outer surfaces of the neuroepithelium, and begins to crawl along it, as if climbing a tree toward the pial surface. As more neurons are born, and more neural stem cells leave the cell cycle, there are fewer and fewer of these poles to climb. By the time a human baby is born, the ventricular zone has been emptied, and the neurons are all up near the pial surface, which has become the cerebral cortex.

The cerebral cortex of mammals is divided into cellular layers. Each layer contains its own types of neurons with particular shapes and functions. In 1961, Richard Sidman, who was working at the U.S. National Institutes of Health, wanted to know whether the different layers of the cerebral cortex were composed of neurons that were born at different stages of development, just as a geologist might want to know the ages of different strata of rocks. To study this, he injected pregnant mice at different stages of gestation with a small amount of a nucleotide (one of the four DNA building blocks) that contained a radioactive isotope of hydrogen, which he used to date the time of cellular birth. The radioactive isotope becomes incorporated into the DNA of dividing progenitor cells during the second phase of the cell cycle. Any cells that were born before the radioactive nucleotide was available remain free of the label. Cells that went through many divisions after the radioactive pulse was available have diluted the label by a factor of two at each division. So, only those cells that went through their very last round of DNA replication at the time of injection retain the highest amount of radioactivity in their DNA.

Sidman dipped the microscopic sections of the isotope-labeled brains into photographic emulsion. The radioactive decay from the incorporated isotope exposed the emulsion, as light would, revealing neuronal birthdates. Remarkably, Sidman discovered that the layers of the cerebral cortex were arranged in a manner that is highly reminiscent of geologic strata. The oldest neurons (those with the earliest birthdates) were in the deepest layers, and the youngest neurons were in the upper layers.[10] Further studies showed that this inside-out pattern is orchestrated in the following way. The first-born neurons stop when they arrive near the pial surface. The next-born cells push past these earlier ones, and then they stop migrating. Each new cohort of new neurons migrates past the preceding ones, and so the "inside-out" architecture of the cortical layers is generated.

A mutant mouse called "reeler" because of its lurching and unstable gait helped us gain insight into the mechanisms that create this inside-out layering. Amazingly, birth-dating studies on the cerebral cortex of reeler mutant mice indicated that cortical layering is largely inverted in these mutants. In reeler mice, the oldest neurons are in the outer or more superficial layers, and the youngest neurons are in the inner or deeper layers, exactly reversed from normal mice. The reeler gene was cloned and discovered to code for a secreted protein, appropriately named "reelin."[11] Reelin is normally made only by the earliest neurons to arrive in the cortex. These reelin-secreting neurons remain close to the pial surface, where they play a role in building the cortex, and they die as soon as their job is done (see chapter 7). In the absence of reelin (i.e., in reeler mutants), the neurons migrate toward the pial surface as usual, but they start piling up onto others that have already migrated, rather than pushing through these previously generated neurons to get to the reelin-rich pial surface. As a result, the older neurons are at the surface, and young neurons are below in the mutant mice. The effects of mutations in the reeler gene can be devastating in humans, as they are characterized by movement disabilities, cognitive deficits, poor language development, schizophrenia, and autism.

The Replacement of Neurons

The cells comprising most of our organs are replaced throughout our lifetimes. Red blood cells survive for about four months and are replaced by new ones. In a typical adult, more than 2 million new red blood cells are made every second. Skin cells are also continuously being born and replaced. They live for just weeks. Cells of the large intestine are replaced every few days.

If injuries occur, these tissues may speed up their proliferation to replace the lost tissue more quickly. Yet the production of neurons in the human brain is complete at birth or shortly thereafter, and if we suffer a brain injury and some neurons die, they are not replaced.

Some lucky animals can replace lost neurons. A flatworm can regenerate a whole new brain! Fish and salamanders are also capable of regenerating damaged parts of their brains and can even replace specific types of neurons when needed. This regenerative ability is lost in the mammalian brain. No one knows exactly how or why this happened.

Some of the strongest evidence that humans stop making new neurons comes from the testing of nuclear weapons. The most common and stable isotope of carbon on earth is ^{12}C, but there is also a little bit of ^{14}C in the atmosphere due to ionizing radiation from space. Live plants and animals take up atmospheric ^{14}C just as they would ^{12}C. But once an animal cell or plant cell dies, the ratio of ^{14}C to ^{12}C decreases as the ^{14}C starts to decay. The half-life of ^{14}C is 5,730 years, so if one does an analysis of the rings of very ancient trees, those rings will have a ratio of $^{14}C / ^{12}C$ that is lower than when they were first formed, which allows us to estimate the age of tree rings by carbon dating. In the early 1950s, countries began testing nuclear weapons above ground. This released large amounts of ^{14}C into the atmosphere, and levels of ^{14}C rose accordingly in the DNA of all cells that were going through proliferation at the time. In 1963, the Partial Nuclear Test Ban Treaty was signed, and since then, nuclear testing has been done underground. As a result, the ^{14}C that was high in the atmosphere in 1963 has been dropping back down toward pre–Cold War levels as it is taken up by trees and fixed in various ways. This "bomb pulse" of ^{14}C, which peaked in 1963, allowed a new kind of carbon dating that can be used to estimate when cells in humans were born. A person born before 1950 would have low levels of ^{14}C in their DNA as a baby, but if that person made any new cells in the 1960s, those cells would have higher levels of ^{14}C in their DNA.

Jonas Frisén, working at the Karolinska Institute in Sweden, has been dating the time of origin of cells in humans, using the changing atmospheric levels of ^{14}C levels created by the bomb pulse. Because blood cells and intestinal cells are replaced regularly, the ^{14}C content of their DNA always reflect the ^{14}C in the atmosphere near the time of the person's age. However, neurons of the cerebral cortex had ^{14}C contents in their DNA that reflected the level of atmospheric ^{14}C at the time of the person's birth. A person who was born near the peak of the bomb pulse and died many years later still had levels of ^{14}C in their cortical neurons that matched the atmospheric level at their date of birth, while the cortical neurons of a person who was born before aboveground testing began, even though they lived through times of high atmospheric ^{14}C, still had low levels of ^{14}C in their cortical neurons, indicative of the low level in the atmosphere at the time of their (and their neurons') birth. Such studies indicate that cortical neurons are not replaced throughout life, and that cortical neurons in a human are as old as the human is.[12]

Stem Cell Niches

Many fish, amphibians, and reptiles continue to grow throughout their adult lives, and their brains grow with them. In these animals, new neurons arise from neural stem cells that reside in specialized local microenvironments in the brain called "stem cell niches." The edge of the retina in a growing fish is a good example of a stem cell niche.[13] Here, new neurons are added every day of a growing fish's life. Some fish can grow for many years, and their eyes grow with them by adding rings of cells at the growing edge, like concentric tree rings. Similarly, the brains of fish, amphibians, and reptiles continue to grow and add new neurons from neural stem cells niches well into adulthood.

While many cold-blooded vertebrates, like fish, may grow substantially as adults, warm-blooded vertebrates, birds, and mammals tend to stop growing as adults. Precious little new neuron generation occurs in the brains of birds. But some does occur, as Fernando Nottebaum at Rockefeller University discovered during his studies of song learning in zebra finches. He saw that in a region of their brains that is specifically involved in song production, these songbirds lose neurons in the off-season and replace them with new neurons in the singing season. It is an annual cycle. It is thought that these new neurons are used to keep up with the seasonal changes in songs. This kind of replacement, Nottebaum suggests, can also be thought of as a process of brain "rejuvenation."

What about mammals? Do mammals have any neural stem cell niches? In 1962, Joseph Altman working at MIT used a cell birth-dating technique on rats and found some newly born neurons in adult rat brains.[14] No one paid these results much mind, because similar studies had found no evidence of new neurons in the cerebral cortex. As a result, Altman's findings were almost forgotten until the 1990s, when several different laboratories discovered two neural stem cell niches in the rodent brain. The first neural stem cell niche supplies new neurons to the olfactory bulb at the front of the brain, where odor signals are first processed. Evidence suggests that this adult stem cell niche is not active in humans. The second niche supplies new neurons to a part of a forebrain area known as the hippocampus. The hippocampus is famous for its role in the formation of memories. A brain scan study of London taxi drivers who were in the process of acquiring "The Knowledge" (a detailed understanding of London's streets) showed that during this training period, their hippocampi seemed to grow. An enriched environment or physical exercise can increase new neuron production in the hippocampi of rats and mice, and ^{14}C bomb-pulse dating in human brains suggests that adult humans appear to make several hundred new neurons every day in the hippocampus. Postmortem analysis of the brains of cancer patients who volunteered to have an intravenous injection of an indelible nucleotide marker of cell proliferation for diagnostic purposes also provides evidence for the generation of new neurons in the adult human hippocampus. However, because nucleotides may also be incorporated into nondividing cells, for example when DNA damage is repaired, there is a concern that all these results could be false positives. Indeed, several other studies, using markers of division, failed to find solid evidence for any neural replacement at all in the hippocampus of adult humans. The results are similar in macaque monkeys, where clear evidence is found for the production of new neurons in the hippocampus of juveniles, but this production declines to undetectable levels in adults. So, some uncertainty still exists in the field about whether we humans produce any new neurons at all in our adult brains.[15]

The connection between new neuron production, if it does occur, and memory formation in the hippocampus is intriguing. One might wonder why the hippocampus would need new neurons to form new memories. One idea that is reminiscent of song learning in birds is that these new neurons hold parts of recent experiences. Some recent experiences are then transferred into long-term memories that are stored elsewhere in the brain, but in the hippocampus, when the neurons holding remnants of the oldest of these "recent" experiences die, it effectively erases any memories or parts of memories that have not made it into long-term storage elsewhere. The dying neurons are then replaced by new neurons that will be exposed to and molded by their own recent experiences and help to build new memories.

We have seen that the human brain grows to its comparatively large size through the proliferation of an initially small population of neural stem cells. At first, these cells divide symmetrically, and their numbers increase exponentially. Subsequently they switch to an asymmetric mode of division, and growth becomes more linear, and then, as birth approaches, they make their terminal divisions. By the time of birth, almost all of them stop dividing. Once born, neurons differentiate into an enormous array of types, each of which is specialized for processing and transmitting particular types of information—the topic of chapter 4.

4

Butterflies of the Soul

In which neurons begin to acquire anatomical and
physiological characteristics marking them as
specific types of neurons that do particular
information-processing tasks in the brain, and in
which we come to see how the specific identity of a
neuron stems from its ancestral nature and lineage,
its exposure to external influences, and chance.

A Cell Becomes a Neuron

The first neurons are born in the human fetus at about week ten
of gestation. When a neuron is first born, it is useless as a pro-
cessor of information. But when it settles into its permanent
position in the brain, it sprouts branches that become dendrites
to take in information, and it grows an axon to send out information. The intricacies of the neuron's branching patterns identify
it as a particular type of neuron. There are at least thousands,
possibly millions, or even billions of different types of neurons.
No one really knows how many types of neurons there are in the
brain, because neuroscientists are still figuring out how to cate-
gorize them.[1] We do know, however, that certain neuron types

are involved in particular aspects of brain function. A very familiar example is the different types of cone cells in our retinas, which are responsible for color vision. Another familiar example is a class of dopamine-secreting neurons in the ventral midbrain, whose degeneration leads to Parkinson's disease. Less familiar are the neurons in the hypothalamus that regulate the sleep cycle by releasing a neurotransmitter known as hypocretin. Loss of these neurons results in narcolepsy. One might think of the brain as a country where no one is out of work, and everyone has a specific job to do. We know a bit about how people get jobs, but how do neurons get their jobs?

Insight into the cell's life as a young neuron came first from partially bald fruit flies discovered in Morgan's genetics lab at Columbia University (see chapter 2). Flies sense touch and wind patterns through the movements of their bristles, each of which is associated with a peripheral sensory neuron. Genetic mapping of mutations in fruit flies in which some or other of the bristles do not form revealed a location in a small region of one of the fly's chromosomes where the relevant gene or genes must be. Antonio Garcia-Bellido and his associates working in Madrid in the 1970s then found another mutant at the same position. But this mutant seemed very different. It was nonviable. In other words, this mutation caused embryonic death: the mutant larvae never hatched out of their egg cases, whereas the bald mutants that lacked bristles were all viable.

To study this lethal mutation further, Garcia-Bellido used a genetic technique to make mosaic flies: animals formed from a patchwork of mutant and normal tissues. For example, the right wing might be mutant and the left wing normal. When he made mosaic flies that were part mutant for the new lethal mutation and part normal, some embryos died and some survived. This result is because of the great variability in where the dividing lines are drawn between mutant and normal tissue using this technique. The mosaic flies that survived had some parts that were mutant, and surprisingly, the mutant parts had normal bristles. But what was the composition of parts for the mosaic flies that died as embryos? Garcia-Bellido and colleagues performed autopsies of hundreds of partly normal and partly mutant fly embryos, to determine which parts of the embryo might be essential for early survival. They found that all the nonviable mosaic embryos had patches of mutant tissue in the same ventral positions. As previous work showed that the central nervous system of the fly comes from such a ventral position (see chapter 2), it was proposed that the new mutant affected the development of the central nervous system, just as the bristle mutants affected the development of the peripheral nervous system.[2] This idea was confirmed by direct examination of the mutant embryos, showing widespread disruption and cell death in the developing central nervous system.

While Garcia-Bellido and colleagues were homing in on these genes, now known as "proneural" genes, Harold Weintraub, working at the University of Washington, was looking for genes that had similar roles in the development of muscle. He started with a standard laboratory line of cells called "fibroblasts," which are cells of the connective tissue. Weintraub introduced genes that were active in muscle cells into these fibroblasts, in what is called a "misexpression experiment." One of the genes that Weintraub misexpressed caused these fibroblasts to turn into skeletal muscle cells. Muscle cells look nothing like fibroblasts, which are star shaped. Muscle cells are long tube-shaped structures full of organized bundles of actin and myosin, the proteins that cause muscles to contract. The transformed fibroblasts not only looked like muscle cells, but they also twitched in the culture dish, just like real muscle cells. This miraculous cellular transformation, from fibroblast to muscle, was controlled by the action of a single gene, a master regulator of muscle development.[3]

The protein made by this gene is a transcription factor that unleashes an entire repertoire of molecular pathways that direct muscle development. When the genetic sequence of the proneural genes was compared with that of this promuscle gene, it became clear that they coded for structurally very similar transcription factors. Promuscle and proneural transcription factors are found throughout the animal kingdom. From flies to humans, the transformation of cells to neuronal fates is marked by the activity of evolutionarily related proneural transcription factors. Like the promuscle transcription factors, the proneural transcription factors turn on many other genes, many of which themselves produce transcription factors that turn on yet more genes. It is this orchestrated cascade that eventually engages thousands of genes essential for the development of a neuron. Neurological diseases caused by mutations that inhibit proneural genes are relatively rare possibly because these genes are as vital for humans as they are for flies.

To Be a Neuron or Not(ch) to Be a Neuron

Every neural stem in the embryo has the capacity to generate neurons, but doing so too early would deplete the neural stem cell population and thus lead to the underproduction of neurons. When a cell is ready to leave the cell cycle and become a neuron, it is because the level of proneural transcription factors has reached a threshold that overcomes the drive to stay in the cell cycle. Once this threshold is reached, it is quickly amplified because each proneural transcription factor not only activates many other target genes but it also activates itself and other proneural genes. Through this mutual and self-supportive gene activity pattern, sufficient amounts of proneural transcription factors are made to engage all the necessary target genes that allow the cell to achieve its cellular fate as a neuron. But at the earliest stages of this process, before the proneural transcription factors have reached the threshold that makes commitment to a neuronal fate an inevitability, the proneural pathway can be shut down. And it often is.

In fact, too many cells try to become neurons, and some must be stopped! Throughout the entire period of generating neurons, battles are constantly waged between nearest neighbors, who are often sister cells. Neighbors compete to keep their own proneural genes turned on and to turn off the proneural genes in their competitors. The weapons used in this competition are components of a molecular pathway named after another mutant called "Notch." What is astonishing is that the effect of Notch pathway mutants is almost opposite to that of the proneural mutants, though both scenarios lead to dead embryos. A deletion of the proneural genes leads to an embryo that lacks neurons, whereas a deletion of the Notch gene leads to an embryo in which far too many cells become neurons and not enough take on non-neural fates. The Notch mutant embryo dies looking like a giant blob of neural tissue with no skin surrounding it.

Laboratories around the world began to try to figure out how the Notch pathway works. Notch is a protein that sits in the outer membrane of the cell. It has an intracellular and an extracellular part. The extracellular part is a receptor that binds to a part of another protein, known as a Notch ligand, which is found on the surface of neighboring cells. When this ligand binds, it exposes Notch to protein-cutting enzymes. One enzyme cuts off the extracellular bit of Notch, and another enzyme cuts off the intracellular bit, freeing these pieces of protein from the cell membrane. The freed intracellular pieces of Notch drift through the cytoplasm and then enter the cell nucleus, where they act as transcription factors, turning on or off other genes. In fact, major targets of the genes that Notch turns on are those responsible for turning off the proneural genes. This seems straightforward enough, but then comes a twist that turns this simple tale into an extraordinary one.

Proneural genes also turn on the ligand for Notch. This completes a positive feedback loop between neighboring cells that are beginning to battle with each other for the right to become a neuron. Cells that have more active proneural genes produce more of the ligand and so are better able to inhibit their neighbors through their Notch pathway. Cells with more active Notch are not only inhibited from becoming neurons, but because of their lower levels of ligand they are also less capable of preventing their neighbors from doing so. This positive feedback loop amplifies very small differences in Notch signaling between neighboring cells. If a group of cells start with roughly equivalent levels of proneural gene activity and Notch signaling, it takes only the slightest imbalance to activate the proneural genes in one cell and completely shut them off in all its nearest neighbors. As we shall see in chapter 7, this competition is the first of many head-to-head battles that a young neuron must win if it is to become part of the brain.

Ramón y Cajal and the Individuality of Neurons

Santiago Ramón y Cajal was born in 1852 in Petilla de Aragón, a village in rural Spain. In his wonderful memoirs, he talks about his childhood love of drawing and painting, and how he wished to become an artist.[4] He was a clever and insubordinate

child, almost expelled from school for drawing a cruel carica-
ture of one of his teachers on the village wall. He tells of how
his father took his artwork to be evaluated by a professional
artist friend of his. This artist gave the verdict that, though he
certainly had ability, it was unlikely that the young Cajal would
be able to earn a living as an artist. With this news, his father
encouraged him to become a doctor. Reluctantly, Cajal agreed
to go medical school. After graduating, he joined an expedition
to Cuba as a military doctor, but he returned to Spain only
months later, desperately ill with malaria and tuberculosis.
Though the illness did not kill Cajal, it ended his military and
medical career in one stroke. When he eventually recovered,
Cajal opted for a career as a histologist, one of his strengths at
medical school, because it involved a lot of drawing. At that
time, very little was known about how the brain worked, but by
the time he died in 1934, Cajal had cemented his place in history
as the father of modern neuroscience.

It had been more than a century since the famous electrical
experimenter, Luigi Galvani, had shown that giving an electric
shock to the sciatic nerve of a frog would cause its leg to twitch.
But still little was known about how the nervous system generated its own electricity or how reflexes worked. The anatomists
and histologists working in the 1800s had seen that neurons have
microscopic, thread-like extensions emanating from their cell
bodies and that some of these fine extensions run into nerves or
the white matter of the brain; they tried to tease apart nerves
into their component microscopic threads, but it was impossible
at that time to follow these extensions for any distance and to
see what they did at their far ends. The brain is full of these
thread-like processes, which emanate from all over and seem to
weave themselves together. It was proposed that the finest of
these extensions fused with one another and that it was this

fusion that allowed electric currents to flow from one neuron to the next. This idea of a continuous meshwork of fused neurons was called the "reticular theory of the nervous system," and it was championed by the great Italian histologist Camillo Golgi.

Many began to doubt the reticular theory in the 1890s because the English physiologist Charles Sherrington demonstrated that the reticular theory of the nervous system seemed incompatible with some fundamental facts about reflexes.[5] Sherrington showed that in most reflexes, when a muscle contracts in response to a sensory stimulus, its antagonistic muscle relaxes. For example, a stimulus that excites the motor neurons that innervate flexor muscles tends to inhibit motor neurons that innervate extensor muscles. This simple fact means that when a sensory signal enters the spinal cord, it leads to the excitation of some motor neurons and the inhibition of others. This implies that neurons communicate using a kind of "plus or minus" logic. Indeed, we now know that neural circuits are made using such principles. One neuron may excite another, which inhibits a third, and so forth. This kind of logic is not easily compatible with the reticular theory, where any change in membrane voltage would seem to be preserved in sign as it flowed from one neuron to the next. Sherrington proposed instead that neurons communicated through specialized junctions which he called "synapses" and that synapses could either be excitatory or inhibitory.

Then along came Cajal, who burned with ambition and curiosity. He not only wanted to resolve the debate about the reticular vs. the synaptic hypothesis, he also wanted to explore the brain, to find out what kinds of cells it was made of. He wanted to know much more. He wanted to know how this lump of tissue was capable of thought. But all these goals depended on whether he could find a way to see what individual neurons looked like

in detail, including all their finest extensions. Cajal began searching for different ways to stain neurons in pieces of neural tissue
that he removed from animals and human cadavers. He revived
and improved a staining method that had been invented and then
abandoned by Camillo Golgi. The Golgi method, as it is now
known, works by infusing a silver solution into the tissue and
chemically inducing its crystallization in neurons so that every
corner of the stained neurons turns dark. The technique is delicate, laborious, and very capricious. Sometimes it does not work
at all, sometimes it stains everything black, sometimes it stains
only a few neurons black, and it is always the case that which
types of neurons it stains is uncontrollable and unpredictable.

When the Golgi technique worked to label just a few cells,
Cajal was mesmerized by the details it revealed about individual neurons "in which the finest branchlets . . . stand out with
unsurpassable clarity upon a transparent yellow background."
It was the capriciousness of the method that caused Golgi to
abandon it, whereas Cajal saw potential of this feature of the
Golgi method as a route that would allow him to see and draw
individual neurons in their entirety. This was exactly what he
needed for his great quests. He devoted his life to fine-tuning
and using the Golgi method to examine the detailed anatomy
of neurons in the brains of all kinds of animals and humans. His
major work, *Histologie du Système Nerveux de l'Homme et des
Vertébrés* (*Histology of the Nervous System of Man and Vertebrates*), published in 1909, is heavily used by neuroscientists
today because many of his observations and detailed drawings
of different types of neurons are unsurpassed and are still
hugely, scientifically useful.[6]

Cajal identified many hundreds of types of neurons that were
heavily stained all along their finest branches, which ended
abruptly at sites of contacts with other neurons. This was

compelling evidence against the reticular theory, which, largely because of Cajal's work, was replaced in favor of our current view of the nervous system, known as the neuron doctrine. The neuron doctrine posits simply that each neuron is an individual cell and that it communicates with other cells through specialized junctions, the synapses that Sherrington predicted.

Golgi, however, did not trust Cajal's results and held on to the reticular theory. In 1906, they shared the Nobel Prize as bitter rivals. It was not until the 1950s that high-resolution electron microscopy finally settled the issue by providing images of the cellular membranes of neurons, showing that each neuron is fully surrounded by its own cell membrane. The images also showed multitudes of synapses, just as Sherrington envisaged. Even Golgi would have had to relinquish the reticular theory in the face of such evidence, but he had died in 1926.

Cajal's efforts to understand neurons inspired the portrait-artist in him. He made stunningly accurate and beautiful portraits of individual neurons, which are now shown in art galleries throughout the world (figure 4.1). In his memoirs, Cajal said, "I must have done over 12,000 drawings. To the layman, they look like strange drawings, with details that measure thousandths of a millimetre, but they reveal the mysterious worlds of the architecture of the brain. . . . Like the entomologist in search of colorful butterflies, my attention has chased in the gardens of the grey matter cells with delicate and elegant shapes, the mysterious butterflies of the soul, whose beating of wings may one day reveal to us the secrets of the mind."[7]

Cajal would not have been able to tell from looking at the anatomy of a neuron how it might also differ from others in the way it functioned. For example, Cajal could not have known which neurons released excitatory neurotransmitters at their

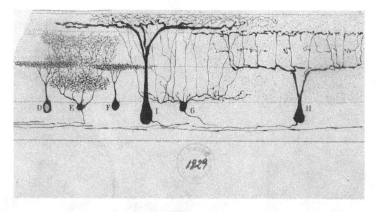

FIGURE 4.1. Cajal's drawing of some different types of retinal ganglion cells. These are from a cross section of a mammalian retina. Courtesy of the Cajal Institute, "Cajal Legacy," Spanish National Research Council (CSIC), Madrid, Spain.

synapses and which released inhibitory neurotransmitters. Physiological and molecular studies since the time of Cajal have substantially expanded our knowledge of cell types. The Allen Institute of Brain Science in Seattle and many other laboratories around the world are working together to construct a full table of neuron types based on an analysis of cell types that includes anatomy, function, and molecular characterization. For example, figure 4.2 shows just one subtype of retinal ganglion cell found in the laboratory of Josh Sanes at Harvard University, one of the many dozens of different subtypes in the retina of a mouse. This one is specifically sensitive to objects that move downward in the visual field, and it is very striking to see that the dendrites of these neurons have all elaborated in the downward direction. This anatomical feature of the neuron's anatomy, which is clearly a key to its sensitivity to downward motion, begins to develop in utero, long before vision is possible.[8]

100 µm

FIGURE 4.2. A specific type of retinal ganglion cell in the mouse retina. This particular type of retinal ganglion cell is sensitive to downward motion. Note that the dendrites of these neurons, when the retina is seen en face, are all oriented downward, revealing a clear relationship between the anatomy of the neuron and its function. From J. Liu and J. R. Sanes. 2017. "Cellular and Molecular Analysis of Dendritic Morphogenesis in a Retinal Cell Type That Senses Color Contrast and Ventral Motion." *J Neurosci* 37: 12247–62. (Photo courtesy of Josh Sanes.)

Cellular Nature and Cellular Nurture

The late Sydney Brenner, Nobel laureate for his work on molecular and developmental biology, contrasted two distinct strategies for achieving cell type identity. The first was based on lineage; Brenner called it the "European Plan, where what matters most is who your ancestors are." The second strategy, based on location, he called the "American Plan, where what matters most is who your neighbors are."[9] Brenner worked on the tiny nematode worm *C. elegans*, which has only 959 cells in its entire body as an adult. Brenner and colleagues recorded the entire lineage histories of all adult cells in the animal back to the single fertilized egg (see chapter 4). They found that the ancestry of almost every cell is the same from animal to animal, suggesting that lineage is likely to be the dominant influence on cellular destiny. Brenner would say that cell type identity in this species is achieved according to the European Plan.

Lineage-dependent mechanisms must involve the passing down of some influence from a mother cell to a daughter cell. But what can a mother cell give a daughter cell to influence the daughter's fate? The nature of one such inherited cellular influence was revealed from the exploration of some mutants of *C. elegans*. If you tickle a nematode on its tail, it squirms forward; tickle it on its head, and it wriggles backward. Each *C. elegans* nematode has just six touch-sensitive neurons along its body. These neurons send signals to other neurons, including motor neurons, that determine the direction of motion. Martin Chalfie at Columbia University searched for mutants that did not respond to tickles. He identified about a dozen genes that were involved in the development of the touch-sensitive neurons. The lineage of each of these touch-sensitive neurons is basically the same. Each one is the great granddaughter of a set of six cells called "Q" that always divides in a way that produces an anterior and a posterior daughter. It is the posterior daughter who is always the grandmother of the touch cell. In one of the first touch-insensitive mutants that Chalfie found, the Q cells produce an anterior daughter as normal; but instead of becoming the grandmother of a touch-sensitive neuron, the posterior daughter becomes a version of her own mother. This posterior daughter cell then divides again to produce another anterior daughter and another copy of itself, and so on. Thus, the grandmother of the touch-sensitive neuron is never generated in this mutant, which, of course, means that the touch-sensitive neurons are never born. Another mutant affects the mother of the touch cell, which normally generates two different types of neurons, only one of which is touch sensitive. In this mutant, the mother cell produces two of the other neurons instead of one of each, so again, touch-sensitive neurons are never made.

Molecular analysis of the genes affected by these two mutants shows that they both make transcription factors. The first of these transcription factors becomes active in the great grandmother. This transcription factor activates a second gene, which comes on later in the lineage, in the mother of the touch-sensitive neuron. When a touch-sensitive neuron is finally born in a normal animal, it has both transcription factors working together to turn on the genes that are needed for its further differentiation as a touch-sensitive neuron. Such lineage-based schemes, in which cells and their daughters accumulate transcription factors through a cascade and then these transcription factors work together to affect cell identity choices at discrete steps of the lineage tree, seem to apply to all lineages in the nematode.

If all neurons in a nematode's nervous system come from Europe, then all neurons in a fruit fly's eye must come from America. Like that of all insects, the eye of a fly is composed of hundreds of facets, each itself a tiny eye with its own little focusing lens. Under each of the hundreds of tiny lenses in a fly's eye is a mini-retina, consisting of eight different photoreceptive neurons and several different types of other cells. In a European Plan, it would seem to make sense that all eight photoreceptive neurons come from a single great-grandmother cell. This, however, is not the case. Any newborn cell in the fly retina can assume a number of different cellular destinies, and the choice is completely wide open, without regard to the cell's lineage and uncorrelated with whatever cell type its sister chooses to become. In a developing fly's eye, a mother cell appears to have no influence at all on the fate of her daughters.

How then is the fly's eye assembled? All the newly born cells are initially equivalent, but through the process of the Notch-mediated inhibition of surrounding cells (as discussed earlier in this chapter), regularly spaced cells are singled out as the first to become neurons. They crank up their proneural genes. Each of these first neuronal cells will become the founder of a small cluster of cells that make a single facet of the eye. A cluster is formed when the founding cell invites its nearest neighbors to join the cluster in an orderly way. As the cells join the cluster, they acquire their specific cellular destinies. Each cell that joins the cluster in a specific position receives signals from its neighbors that previously joined the cluster. These signals turn on combinations of transcription factors that determine which type of eye cell to become. This process has been likened to how a crystal grows, and the fly's eye can indeed be considered as a kind of neurocrystal, formed of cells that achieve their fate not through lineage but by recruitment into a self-organizing tissue.[10] This wave of cellular crystallization sweeps across the developing fly eye like a wave heading into a sea of cellular chaos and leaving in its wake an organized array of developing mini-eyes (figure 4.3).

The nematode central nervous system and the fly retina are rather extreme cases of "European and American Plans." Lineage dominates the first, and cellular environment dominates the second. These examples also demonstrate what might be called "cellular nature" and "cellular nurture." It turns out that most neurons, especially those in larger brains, use a complex combination of intrinsic and extrinsic influences to reach their

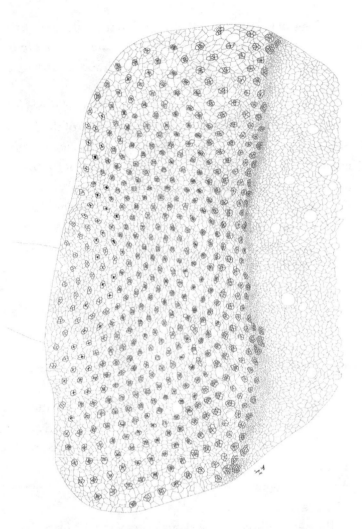

FIGURE 4.3. The formation of the *Drosophila* neurocrystalline retina. On the right, the cells seem disorganized; on the left, the cells have organized themselves into clusters that will become the facets of an adult retina. Near the center, two-thirds of the way across the retina, one can see the dividing line between the disorganized and the organized cells as the wave of crystallization sweeps across the retina from left to right. (This image, drawn by Tanya Wolff, is from a poster insert and is included in Michael Bate and Alfonso Martinez Arias (eds.). 1993. *The Development of Drosophila melanogaster.* New York: Cold Harbor Laboratory Press.)

final cellular destinies. An extrinsic signal may lead to the activation of a transcription factor, which becomes an intrinsic feature of the cell. This transcription factor may allow this cell to take the next step on the pathway to its final destiny by turning on genes that code for a receptor that makes the cell responsive to a new signal. If the cell receives this signal, it turns on a new transcription factor, and so on.

An incredible example of how neurons in the spinal cord reach a decision about which type and subtype of neuron to become comes from the study of motor neurons in the spinal cord.[11] Tom Jessell and his colleagues at Columbia University showed that in response to sonic hedgehog, cells in a ventral domain of the neural tube activate transcription factors that turn these cells into motor neurons. But the transcription factors they turn on in response to sonic hedgehog are in all motor neurons. These are "generic" motor neurons that have not yet been branded with the details of their fates, e.g., which of the 600 or so muscles in the body they are meant to innervate. The subdivision of motor neurons into different types is a step-by-step process. First the subdivision is arranged according to "motor columns." The motor neurons that innervate the head, the arms, the trunk, and the legs are arranged in separate columns along the length of the spinal cord by the action of Hox transcription factors (see chapter 2), which work together with generic motor neuron transcription factors to give motor neurons columnar identities. Motor columns are further divided into those that innervate flexors and those that innervate extensors. This subdivision is the result of yet other transcription factors that regulate that choice. Finally, motor neurons become committed to innervate specific individual muscles. This constitutes the highest level of motor neuron subtype classification, known as "motor pool" identity. Motor pool identity is, of course, the result of turning on yet other transcription factors.

Jessell and colleagues identified many components of the signaling pathways and transcriptional cascades that end in the expression of motor pool transcription factors. An example of one such motor pool transcription factor is Pea3, which is expressed in motor neurons that innervate the shoulder muscle known as the latissimus dorsi. The latissimus dorsi is used for forward movement in quadrupeds (e.g., mice and horses) and for adducting (bringing the arm into the body) in bipeds (e.g., humans and birds). The motor neurons of the latissimus dorsi motor pool do not develop in Pea3 mutant mice. As a result, the latissimus dorsi muscles atrophy due to lack of innervation, and the mice have problems walking and running.

The work of Jessell and colleagues on vertebrate motor neuron types lays out a type of molecular logic by which many neurons acquire their specific destinies. The accumulation of transcription factors takes place through an interaction of lineage *and* the cellular environment. The understanding of such logic can be used by scientists to grow embryonic stem cells in tissue culture and turn them into particular types of cells in the body, including particular types of neurons. Exposure of embryonic stem cells to anti-BMPs turns them into neural stem cells; giving them the right amount of sonic hedgehog turns them into generic motor neurons; and exposing these motor neurons to the retinoic acid activates Hox genes that push these motor neurons to become particular types of motor neurons. By understanding developmental principles of cell type determination, it is becoming possible to use embryonic stem cells from humans to generate almost any type of cell in the human body or brain for study in culture and research into potential treatments of various diseases, including diseases of the nervous system. Scientists are already using such strategies to search for cures to neuron-type specific, degenerative diseases, such as Parkinson's disease, in a tissue-culture dish.

Encountering Fate and Neuroblastoma

The neurons that populate the peripheral nervous system come from itinerant lineages. These voyagers of the embryonic nervous system acquire their destinies through their travels. The peripheral nervous system arises from the neural crest cells (see chapter 2) that originally occupy the most dorsal regions of the neural tube. Soon after the neural tube closes, neural crest cells begin to migrate out of the neural tube and travel through the body. These neural crest cells proliferate as they migrate, and growing hordes of neural crest cells travel along a myriad of different pathways to their final destinations throughout the far reaches of our bodies. Neural crest cells generate all sorts of cell types. They make smooth muscle, cartilage, bone, teeth, pigment cells, hormone-secreting cells, and the walls of blood vessels; they also make the entire peripheral nervous system. The peripheral nervous system is itself composed of four parts: the sympathetic nervous system ("fight or flight"), the parasympathetic nervous system ("rest, digest, and breed"), the enteric nervous system (which innervates the gut), and the peripheral sensory nervous system.

The first challenge that a neural crest cell faces is to break out of the neural tube. The neural tube is an epithelium in which cells are held together by adhesive molecules, which neural cells use to stick to one another. The neural crest cells loosen these sticky contacts with their neighbors and then begin to enzymatically digest their way through the outer wall of the neural tube, which consists of a heavy matrix of extracellular materials like collagen. Once neural crest cells have escaped the confines of the neural tube, they begin to migrate. In this sense, neural crest cells are like cancer cells that become metastatic, which use similar strategies to break out of their primary tissue and migrate to invade other tissues.

In the late 1960s and early 1970s, at the University of Nantes in France, Nicole Le Douarin devised a simple method to identify all the descendants of migrating neural crest cells. The method was based on her discovery that under the microscope, the cells from quails can easily be distinguished from the cells of chickens. When Le Douarin transplanted quail cells to chicks and vice versa, she could always tell which cells came from the host and which from the donor. After mapping out routes taken and tissues made by neural cells on their migrations, she then wanted to know whether the migrations themselves influenced the fates of the cells. In a classic experiment, she transplanted pieces of premigratory neural crest from the neck region of chick embryos into the trunk region of quail embryos. She found that the donor neck crest cells, if they were transplanted to the trunk region, migrated along trunk routes and differentiated into trunk cell derivatives (i.e., sensory and sympathetic neurons). Vice versa, when premigratory neural crest cells from the trunk were transplanted to the neck region, the progenitors migrated along routes normally taken by neural crest cells from the neck and became the cell types normally derived from neck neural crest cells (i.e., parasympathetic and enteric neurons). This experiment established that at the beginning of their migration, neural crest cells are free to choose their final cellular destinies based on what they encounter in their travels.[12]

Human diseases arise from defects in neural crest migration and cell type specification. For example, Hirschsprung disease is caused by the failure of neural crest cells of the neck region to migrate to the gut, with the result that the enteric nervous system is not properly generated. Without appropriate neural control, the gut cannot contract and move stool well. The digestive systems of babies born with Hirschsprung disease become blocked and backed up. Surgery is required to remove any portion of the intestine that lacks this innervation. A particularly frightening disease of the neural crest is neuroblastoma, a rare but aggressive malignancy of childhood.[13] Most neuroblastoma tumors are made from neural crest progenitors called "sympathoblasts" that normally give rise to neurons of the sympathetic nervous system, or the adrenaline-secreting cells of the adrenal glands. Instead of fully differentiating, these sympathoblasts continue to proliferate and make tumors anywhere along the pathway from the neck to the pelvis, often before they reach their final destinations and achieve their final fates. Because this cancer often arises in a cell type that is already migratory, it often has metastasized by the time of detection. There is a fine balance between differentiation and proliferation in neural crest cells, and neuroblastoma results when the balance swings too much in favor of proliferation. Rarely without other treatment, the balance shifts back in favor of differentiation, so that tumors spontaneously regress. Once cells differentiate into neurons, they do not divide again, so neuroblastoma, like retinoblastoma (see chapter 3), is a disease of early childhood.

The Fourth Dimension—Time

Clearly, where a cell is located within the three spatial dimensions of the body can affect its fate. But there is a fourth dimension to be reckoned with: time. The question is whether time or order of birth has significant influence on fate. One of the most obvious examples of a temporal dimension to cell fate in the nervous system is the fact that neurons tend to be born before glial cells. This issue has been explored in the peripheral nervous system. When neural crest cells first arrive at a destination that instructs them to become part of the peripheral nervous system, they can become either neurons or glia, but the first cells to arrive always become neurons because this is their intrinsic bias. As soon as they turn into neurons, they begin to secrete a protein that inhibits later-arriving neural crest cells from turning into neurons. As more neurons accumulate, the level of the inhibitor protein builds, and crest cells that are late to the party are exposed to sufficient levels of the inhibitor to prevent them from becoming neurons. They become glial cells instead. A similar transition from neuron to glia (specifically, astrocytes) occurs in neural lineages in the central nervous system.

The idea that cells born at one time can influence the fate of later-born cells suggests a basic mechanism for a developmental clock. Chris Doe and his colleagues at the University of Oregon have gone further and provided evidence for a molecular counting mechanism in the development of the embryonic nervous system of the fruit fly—somewhat like the pendulum in a grandfather clock swiveling an escapement mechanism that allows a gear to ratchet forward with each swing. What is being counted in the neural cell lineage of the fly embryo are not swings of a pendulum but cycles of cell division. The embryonic neuroblasts that Doe and colleagues study always divide asymmetrically to produce one large daughter that remains a neuroblast and a smaller daughter, a secondary progenitor called a "ganglion mother cell," or GMC. Each GMC then divides once to produce two mature neurons. The first ganglion mother cell (GMC1) of a particular neuroblast makes two neurons with distinct but often related individual fates. For example, both daughters may become motor neurons. Later-arising GMCs make other types of neurons. A GMC3 may make two inhibitory interneurons with short axons. As they divide asymmetrically, the neuroblasts go through a fixed sequence of expressing what Doe has called "temporal transcription factors"

(TTFs). Let us call these "TTF1," "TTF2," "TTF3," and so forth (although, of course, each TTF has its own weird biological gene name). During each cell cycle, there is a switch from one TTF to the next TTF. When a neuroblast that is expressing TTF1 divides asymmetrically, its smaller daughter, GMC1, inherits TTF1 expression. The larger daughter remains a neuroblast but replaces TTF1 with TTF2, so its next GMC daughter (GMC2) inherits TTF2. The daughter neuroblast then switches to TTF3, and the neuroblast turns off TTF2 and turns on TTF3, and so on. With each cell division, the count ratchets forward, and the new TTFs cause the generation of new cell types. The mechanism that is responsible for the TTF sequence is linked to the cell cycle and a push forward, because each TTF in the neuroblast promotes the expression of the following TTFs while repressing the expression of previous ones.[14]

The layering of the different cell types in the cerebral cortex (chapter 3) is a great example of the forward march of development and of the irreversibility of traveling too far along a particular road. The deep layers of the cerebral cortex are filled with large neurons that are born early. They send axons to the thalamus, the midbrain, the hindbrain, and the spinal cord. Smaller neurons of the middle layers, which are generated next, send their axons to upper cortical layers. Middle-sized neurons destined for the top layers of the cortex are generated last and send their axons to other cortical areas. In the 1980s, Sue McConnell and colleagues at Stanford University investigated whether the newborn neurons become committed to their specific layers and cortical cell types even before they migrate into position. To do this, McConnell and her team transplanted cortical progenitors across time, what developmental biologists call "heterochronic transplants." McConnell used different stages of neonatal ferrets, which are born very prematurely, when cortical neurons are still being generated. When McConnell transplanted early progenitor cells into the brains of older animals, the cells changed their fate and made upper-layer neurons. This suggested that these young cortical cells are flexible with regard to which specific cortical fate they take. The reverse transplant (older stage to younger stage), however, gave very different results. These transplanted older progenitors were not flexible. They did not change, even though they were surrounded by young progenitors. Similarly, when progenitors at intermediate stages of cortical development were transplanted into older brains, they changed their fates, but when they were transplanted into younger hosts, they did not. These experiments suggest that as cortical stem cells divide, they move through a series of stages, from which cells can travel forward but they cannot go backward.[15]

Chance and Fate

The vertebrate retina is a neural tissue that is particularly well studied because of its beautiful organization into clear layers. It is the bit of the brains of frog and fish embryos that I spent most of my life studying, and this is certainly one of the reasons why there are so many stories about the retina and the visual system in this book. The beautiful organization of the retina also attracted the attention of Cajal, who explored the cellular structure as well as the circuitry of the retina. The retinas of all vertebrate animals are organized into three cellular layers, each composed of specific cell types. The outer layer (farthest from the lens) contains the light-sensitive rod and cone cells. Rod and cone cells make synapses with bipolar cells in the middle layer. Bipolar cells are spindle-shaped neurons that look somewhat similar at both ends, except that one end is the dendrite that takes input from photoreceptors, and the other inner end is the axon that makes synapses with retinal ganglion cells, which sit in the inner layer (closest to the lens). The retinal ganglion cells provide the output of the retina. They send their long axons along the optic nerve into the brain. Birth dating studies showed that different retinal cell types are born within different, but overlapping, temporal windows of development and that, as is true in an insect's central nervous system and cerebral cortex, time runs in only one direction for making fate decisions. At early stages, a retinal stem cell can produce all types of retinal neurons, but as the lineages of these retinal stem cells play out, the cells gradually lose their competence to produce earlier cell types.

In one set of experiments in my lab, we used time-lapse microscopy to follow retinal stem cells in the zebrafish retina through all their divisions and to account for all their descendants. Unlike the case of nematodes, there is substantial variability at the level of single stem cell lineages in the vertebrate retina. It seems that each retinal stem cell produces a unique and somewhat randomized constellation of neuronal descendants. Even when single retinal stem cells are isolated in culture dishes, they still produce variable lineages, suggesting that this variability is an intrinsic feature of retinal stem cells. Though it is clearly the combination of transcription factors that influence retinal cell type, it seems to be a matter of chance whether some of the genes for these transcription factors are turned on or off. And just as with cell number variability (chapter 3), the law of large numbers ensures that when such randomizing influences are involved in the lineages of thousands or more equivalent stem cells, close to the expected numbers of each cell type will be generated in the end, even though each lineage is likely to be unique.

One of the personal highlights of my scientific career was to give the Waddington lecture in 2017 to the British Society of Developmental Biology about this work. Carl Waddington was a champion of developmental biology theory for his amazing conceptual insights into development mechanisms. One of the graphical metaphors he produced to show some of his thinking was a simple representation of how cells chose their specific fates in what Waddington referred to as a developmental landscape.[16] There is a downward-sloping valley with elongated hills that divide the valley into narrow glens. Down this valley rolls a ball. The ball may be a progenitor cell. When the ball encounters the first hill, it chooses the glens to the left or to the right, limiting its potential fates. It makes further left/right fate-limiting choices as it rolls on. This metaphor has captured the interests of developmental biologists for decades. What are these hills and glens? How can they be explained by the molecular geography of embryogenesis, and what exactly is going on inside the cell as it rolls down toward its fate-making commitments? How does a cell choose between left and right? I certainly did not try to answer all these questions in the lecture, but I suggested that Waddington's developmental landscape might also be a good way to understand the effects of random influences in development. Imagine rolling a thousand balls down Waddington's valley and that there is some unpredictability about whether each ball goes left or right at each fork. Though one might not be able to predict exactly where any single ball ends up in such a situation, one could still be confident in the general shape of the distribution of balls, or for the case of cells, the proportions of each type that would be generated (figure 4.4). I am not suggesting that development is a random process, or that cells choose their fates in a totally random way, but any element of variability that does exist among lineages in the central nervous system of animals with large numbers of progenitors should only help ensure that the right numbers of each of the neuronal types are made.

It is now possible to look into a human eye with a powerful ophthalmoscope that allows one to distinguish individual rod

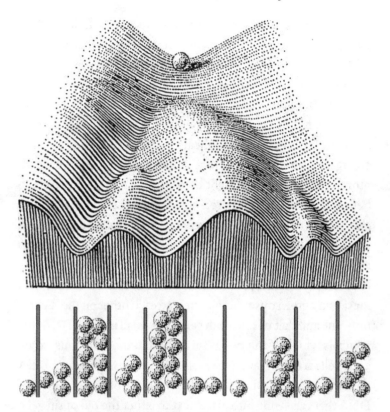

FIGURE 4.4. Waddington's developmental landscape (top) incorporating chance (bottom). Consider a ball rolling into a valley and choosing a path either to the left or the right of the first hill that it encounters and then making a second choice when it encounters the next hill. At the end of the journey, the cell has chosen a specific fate. Many factors may influence whether the ball/cell goes right or left at each choice, including an element of chance. So even though one might not be able to predict the fate of the cell before it commences its journey down the developmental landscape, when many identical balls are rolled into the landscape, they create a predictable distribution of fates by chance alone.

and cone cells at the back of the retina. Here one can see three kinds of cone cells (red, green, and blue), accounting for our ability to see three channels of color information. If one looks closely at how these cells are packed together, one sees that the red and the green cone cells are arranged randomly with respect to each other. This randomness stems from the recent evolutionary origin of trichromacy (three-color vision) in primates, about 30 million years ago. Our more remote ancestors were dichromatic (two-color vision). The genetic events that led to trichromacy involved the duplication of a gene that codes for the protein known as red opsin, the photosensitive protein that captures red light. This gene was duplicated, and then it mutated. The change in the duplicated gene caused the new protein to be more sensitive to green light than to red light. So now, where before there was just one red opsin gene, there are two opsin genes, a red and a green. This genetic event was thought to have spread through our ancestors, because it allowed them to distinguish red from green, and thus ripe from unripe fruit. Jeremy Nathans at Johns Hopkins University discovered that the genes for red and green opsin sit right next to each other on the X chromosome and that next to both genes is a small piece of DNA that controls which of the two is turned on in the cell. Thus, some cone cells are green and others are red. This small piece of DNA that acts as a controller is dangling on the bottom of a loop of DNA that can swing into place next to either the red opsin gene or the green opsin gene, and as the two configurations of the DNA are mutually exclusive, only one of the two genes gets activated in a cell. Finally, the controller seems to choose one or the other by chance, so it is basically a coin toss whether a particular cone cell will be red sensitive or green sensitive.[17] Except for this final choice of which color opsin to express, these two types of cone cell would be classified as the same cell type.

Earlier in this chapter, we mentioned Cajal's fanciful butterflies of the soul, so it seems somehow appropriate to consider the real butterflies of the woodlands. Claude Desplan, working at New York University, has been fascinated by insect color vision. Working first with fruit flies, he found that, as in humans, there are two forms of essentially the same photoreceptor cell type in the fruit fly retina (sensitive to different colors) and that these are arranged randomly such that some of the 800 facets of a fruit fly eye see a different spectrum of colors than the others. It is impossible to predict whether any particular facet will develop one color sensitivity or the other, because the switch is thrown by a chance mechanism, very reminiscent of the red vs. green cone cells of the primate retina. In the fly, the switch has to do with the random turning on or off of a single transcription factor that controls opsin choice. Once the photoreceptor has decided everything else about its identity, it is as though it pulls the lever of a slot machine and gets a result that is either one color or the other. Desplan's attentions were next drawn to swallowtail butterflies, which have more complex eyes. Desplan and colleagues found not one, but two photoreceptor cell types in each facet that pull the lever, and the result is that they are randomly assigned colors independently of one another in each of the thousand facets of their eyes. This allows four different types of facets: ones in which the transcription factors are on in both cells, ones in which they are off in both cells, and ones in which one transcription factor is on and the other is off and vice versa. This random process works with the butterfly's five different colored opsins to allow the animal to do incredibly in-depth color comparisons that far surpass our own abilities. Desplan argues that this simple chance mechanism allows the evolution of expanded color comparisons that give butterflies a supreme ability to recognize flowers, find food, and identify mates.[18]

This unpredictable, yet developmentally useful, aspect of neural development preoccupied me during the last phase of my career in the lab. It initially seemed wild and wonderful that the way brains are made depends on the statistics of chance rather than on a fully determined plan. Yet as we learn more about the dynamics of genes, how they flick on and off, it now seems inevitable that every neural lineage in our brain is likely to be affected by seemingly random molecular events in neural stem cells. Of course, this means that though all human brains have about the same numbers of every cell type, it is extraordinarily unlikely that any two human beings have ever been, or will ever be, born with identical complements of different types of neurons. Every human brain is made in the same way, but every human brain is different.

Neurons as Individuals

In 1998, a gene called "DSCAM" (Down syndrome cell adhesion molecule) was isolated from a region of human chromosome 21.[19] This gene is critical for some of the symptoms of Down syndrome, which is due to possessing three copies of chromosome 21 instead of just two. As the full name indicates, DSCAM is a cell adhesion molecule. It is expressed primarily in the brain during embryonic development, and overexpression of DSCAM also leads to symptoms reminiscent of certain aspects of Down syndrome in mice. Soon after DSCAM was identified in humans, Larry Zipursky at UCLA found a *Drosophila* version of the DSCAM gene. The astonishing thing about the *Drosophila* DSCAM gene is that this single gene produces tens of thousands of distinct proteins.[20] The gene is divided into four coding parts, and each of these parts has a variety of choices: 12 choices for the first part, 48 for the second, 33 for the third, and 2 for the fourth. When the gene is activated, it is first made into messenger RNA (mRNA). The mRNA is then cut into pieces to get rid of noncoding sequences and all but one of each of the coding parts. Because cut-and-splice choices are probabilistic, each spliced mRNA is then translated into one of $12 \times 48 \times 33 \times 2$, or

38,016 possible cell adhesion molecules. This huge diversity of possible proteins from just a single gene is reminiscent of the immune system, where recombining genes coding for different protein domains generates a tremendous diversity of antibodies. Indeed, DSCAM is a member of what is known as the immunoglobulin superfamily of cell adhesion molecules. Cell adhesion molecules in this superfamily have extracellular domains that are related to those used by antibodies to recognize antigens. Though DSCAMs in flies and many other invertebrates do have an immunity function, what is particularly surprising is that a huge fraction of all possible proteins made from the DSCAM gene are expressed in the fly's nervous system during development, and because of the random alternative splicing, almost every neuron in the fly expresses a different version of DSCAM. Thus each neuron should, in theory, be able to distinguish itself from others.

Although vertebrate animals do not generate as much diversity from their DSCAM genes, they more than make up for that with other cell adhesion molecules. For example, the human genome has three groups of genes called "clustered protocadherin genes" that make an immense array of possible cell adhesion molecules.[21] Though the molecular mechanism for generating random combinations is somewhat different, each one of these clusters, like the DSCAM gene in flies, makes a multitude of different mRNAs. When these mRNAs are made into proteins, the result is that every neuron gets a unique set of protocadherin molecules, giving it a molecularly distinctive surface. It is like a random-number generating process that provides a unique barcode and a unique identity to every neuron.

Our unique identities as human beings allow other human beings to distinguish us, and it allows us to distinguish ourselves from others. Our immune systems are tuned in to our individual molecular identities. It seems every human is likely to have a unique combination of what are called "histocompatibility proteins" that tell the immune system whether the cells in our bodies are our own or are from another person, which is why the immune system has to be suppressed in patients who are recipients of organ transplants. But what good can it do for every neuron in our brains to have its own identity? One possibility is that it gives neurons the capacity to distinguish themselves from other neurons of the same subtype, allowing all the fine branches of a single neuron to recognize one another. This turns out to be critical for wiring up the brain. For example, upon running into each other in the meshwork of axons and dendrites, two branches of the same neuron can use their unique barcode to avoid making useless connections with each other. By examining the neurons that are used in the detection of directional movements, Josh Sanes and colleagues at Harvard University have been investigating the significance of such self-avoidance for the visual system of the mouse retina.[22] When Sanes and colleagues altered these neurons so that they did not express their unique protocadherin self-identifiers, they made synapses on themselves rather than on each other. And when they altered them so that all the neurons of this type expressed the same rather than different identifiers, they did not make any connections with one another as they normally do. In each case, the result was a loss of ability to see directional movements. In humans, mutant variants of the protocadherin clusters have recently been linked with schizophrenia.

The brain employs a huge variety of specialized neuron types in the various regional offices in the brain, and each individual cell has its own role to play in that office. The type of job that a newly born neuron is assigned comes through a combination of extrinsic signals, intrinsic cellular biases, order of birth, and random influences. In the end, each neuron becomes a particular type, each with its own intricate and distinctive branching patterns and physiological attributes. Neurons are also assigned unique identities based on a random combination of a vast array of possible cell adhesion molecules that allows individual neurons to distinguish themselves from others of the same type. Yet these neurons have just begun their lives. They "know" who they are now, but they have not yet made the connections with other neurons that allow them to process and transmit information. One of the first tasks of a new neuron is to begin to wire up the brain.

5

Wiring Up

In which young neurons send out axons that navigate through the developing brain in search of specific destinations that are often far away.

Feats of Navigation

Each neuron in your brain receives electrical inputs on its branched dendrites. The neuron does computations based on these inputs, and sends out the results as electrical impulses that travel along its axon to target neurons elsewhere in the brain. These target neurons do their own computations based on their inputs, and so on. By birth, a human baby has about 100 billion neurons! These neurons connect with one another in ways that allow a stupendous amount of information to be processed and acted on. Neurons in your hypothalamus sense hunger, while neurons in your retina are seeing a pattern in the visual scene in front of you; the neurons in your visual cortex interpret the scene as an English muffin fresh from the toaster. Neurons in your olfactory cortex interpret the smells received as butter melting on the muffin. The neurons in your frontal cortex receive these pieces of information, integrate them, and send signals to the motor cortex. The neurons of the motor cortex organize an action sequence and send impulses down their axons into the spinal cord to motor neurons, which send impulses down *their* axons to activate a particular sequence of muscle contractions in your arms and fingers. The smooth execution of such a string of neural events demands a degree of precise wiring, much of which happened in your fetal brain long before you were born. It is because the wiring happened correctly that you can now bring that English muffin to your mouth and have that well-deserved bite.

One of the greatest of all challenges for developmental neuroscience has been—and still is—to understand how the brain wires itself up. How do the axons of neurons seek out and find their target neurons elsewhere in the brain? A simple way to ensure that all the necessary connections are made would be having every neuron make synapses with every other neuron. Indeed, that would be possible in animals with a small number of neurons, but with as many neurons as a human has, the brain would have to be at least a hundred times larger than it is to contain the necessary all-to-all wiring. So that is not how it is done. We also know that neurons do not simply make random connections with one another, because to a first approximation, all human brains are wired up in very similar ways. Instead, it seems that each neuron makes a limited number of incredibly precise connections. A reasonable approximation is that an average neuron in our brain connects to about 100 other neurons, although some types of neurons are much more selective, and others are much less so. The average selectivity is therefore on the order of one target neuron in a billion.

What is mind boggling is that the neurons themselves somehow seem to "know" how to wire up. One might be tempted to imagine tiny Terry, the electrician, hiding inside an unborn baby's brain, consulting a wiring diagram, and plugging cables into the right sockets. But you know that's not how it works! Neurons manage to wire themselves up without Terry's interventions.

Consider a motor neuron in your embryonic spinal cord sending its axon out of the spinal cord and finding its way to a specific muscle in your leg. This is the equivalent, when you scale sizes up, of a person who lives in Frankfurt going to a particular village on the Riviera. Even with a map, most humans would make some wrong turns along the way, yet growing axons rarely go off course. In addition to sending an axon to specific muscles, motor neurons also send out branches from their main axons that navigate to different targets. These secondary axons search out their targets, inhibitory neurons known as Renshaw cells. So, whenever a motor neuron fires an impulse along its axon to activate a muscle, it also excites a few inhibitory neurons (the Renshaw cells). These neurons inhibit other motor neurons, particularly those whose targets are the antagonistic muscles. This little spinal circuit ensures that agonists and antagonists tend not to fight each other, making many actions more efficient. The point here, however, is that a motor neuron has one axon that navigates to a faraway target, a leg muscle, and another axon with a completely different target in the central nervous system.

In the developing fetus, billions of growing axons are traveling simultaneously through the brain and body in sundry directions, heading to destinations near and far. Almost every one of them appears to know where it is going. Traces of these remarkable feats of axonal navigation are reflected in the anatomical weavings of white and gray matter in the adult brain, with complicated names best known to neurosurgeons and neuroanatomists.

The Growth Cone

Histologists of the nineteenth century knew neurons had long axons, but how they got these long axons was still mysterious, as it was not possible at the time to look inside the developing brain and watch it happen. Then in 1907, developmental biologist Ross Granville Harrison, working at Johns Hopkins University, discovered a method "by which the end of a growing nerve could be brought under direct observation while alive."[1] His method was to remove a little piece of tissue from a frog embryo and put it onto a glass coverslip for a microscope slide. Then he put a couple of drops of lymph from an adult frog on top of the piece of embryonic tissue. The lymph clotted into a transparent gel, holding the tissue in place and providing it with some nutrients. Harrison then inverted the coverslip over a hollow microscope slide and sealed the rim with wax. He was then able to look through the coverslip into the tiny piece of tissue, which might stay alive for up to a week or more. In such a situation, Harrison could observe individual cells minute by minute, hour by hour, or day by day (figure 5.1). This pioneering technique of cell biology, known as tissue culture, where cells can be kept alive or grown in a dish or a flask, allowed Harrison to see axons emerging from young neurons. He described the numerous "fibers" that extended from the explant and into the gelled lymph, where they could be seen as individual growing threads of axons. The most remarkable feature of these extending axons was that at the tip of each one was an enlarged ending that was continuously "changing its shape so rapidly that it was difficult to draw the details accurately."[2] The great Santiago Ramón y Cajal (see chapter 4) had seen such enlarged endings at the tips of axons in his sections of the embryonic spinal cord.

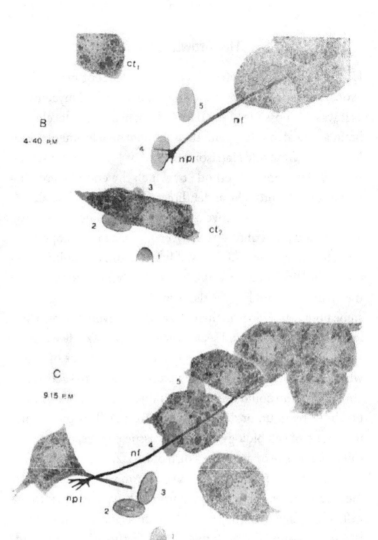

FIGURE 5.1. A growing axon from Harrison's 1910 study. An axon (nf) tipped with growth cone (nPI) at 4 P.M. (top) and 9:15 P.M. (bottom). Note the stable position of red blood cells (about 20 microns along their long axis) numbered 1–5, which serve as markers for how far the axon extends.

In his typically flamboyant style, Cajal wrote, "I had the good fortune to behold for the first time that fantastic ending of the growing axon. In my sections of the spinal cord of the three-day chick embryo, this ending appeared as a concentration of protoplasm of conical form, endowed with amoeboid movements. . . . This curious terminal club, I christened the growth cone."[3] How remarkable must have been Cajal's insight into the nervous system that he intuited the active movements of growth cones from fixed sections, movements that Harrison observed for the first time 20 years later.

Let's look inside this growth cone to see how it works (figure 5.2). The most obvious feature of a growth cone's internal workings is its dynamic cytoskeleton (cytoplasmic skeleton, the tiny molecular cables that give cells structural support). This cytoskeleton is an interconnected framework of extending and shrinking microscopic filaments. One of the most important types of cytoskeletal elements are the microtubules that enter the growth cone from the extending axon. Microtubules are composed of protein subunits called "tubulin." Tubulin subunits are added to just one end of growing microtubules, and these growing ends are all pointed in the same direction as that of the growing axon. As they emerge from the axon into the growth cone, microtubules splay out, helping to give the growth cone its conical shape. The elongation of microtubules helps drive the growth cone forward. In chapter 3, we learned that microtubules also make the spindle that is used to separate duplicated sets of chromosomes in dividing cells during mitosis. This fact spurred the search for drugs that block microtubule assembly for use in chemotherapy to stop the rapid division of cancer cells. When such drugs are applied to growth cones, the axons simply stop growing.

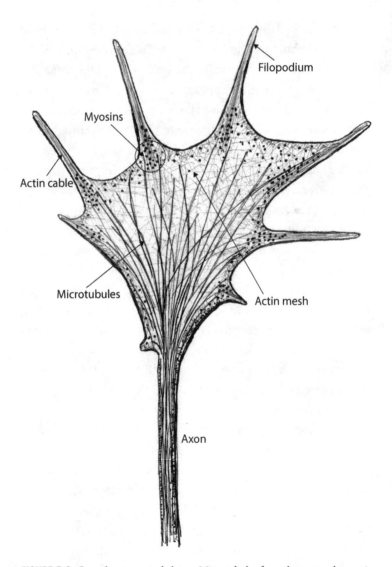

FIGURE 5.2. Growth-cone cytoskeleton. Microtubules from the axon splay out in the central part of the growth cone. Actin cables fill the extending filopodia at the leading edge of the growth cone, and crisscrossing actin filaments fill the peripheral part of the growth cone. Myosin molecules at the base of the filopodia pull on the actin cables, which are linked to the substratum by transmembrane adhesion molecules, causing the growth cone to move forward.

The more dynamic, but no less important, components of the growth-cone cytoskeleton are the actin filaments, which are elongated polymers made of actin subunits. Actin filaments are much thinner and shorter than microtubules. In the center of the growth cone, the actin filaments form an intricately branched meshwork, while at the leading edge of the growth cone, actin fibers bundle together to form thick cables that protrude from the center like fingers of a hand. These actin-filled fingers are called the "filopodia" of a growth cone. If one looks at a live growth cone under a microscope, one can see filopodia shoot out from the front end of a growth cone. They often swing around a little, and then they retract back in, as if the growth cone was somehow using these filopodia to probe for the correct way ahead. Indeed, this appears to be the case in part because if growing axons are experimentally exposed to drugs that inhibit the formation of actin filaments, growth cones lose their filopodia and then often head off course.

Observations in live growth cones show that they use a method of locomotion that is similar in many ways to that of an army tank. Let's start with the actin cables in the filopodia. Each actin cable is oriented so that its growing end is pointed outward, away from the growth-cone center. It continuously adds new subunits of actin molecules at this growing end, lengthening the cable in the forward direction. Yet at the same time, the actin cable is being pulled backward, toward the center of the growth cone, where its non-growing end is being disassembled. The front end grows at about the same rate as the back end gets chewed away. The result is that actin cables in the filopodia remain pretty much the same length, although the actin proteins that compose them are constantly moving rearward like the bottom belt of a tank. The motor of the growth cone is not a diesel engine, but a troupe of myosin molecules, similar to the ones that pull on actin cables in our muscle cells. Situated at the base of the filopodia and attached firmly to the central cytoskeleton of the growth cone, the myosin molecules heave away at the actin cables, continually hauling them in.

With a tank, it is easy to appreciate how traction is generated. The treads on the bottom of the belt engage with the ground, and if they do not slip, then by Newton's third law of motion, by putting backward force on the ground, the whole tank moves forward in the opposite direction. But in the growth cone, the actin cables are inside the cell's membrane, so they themselves cannot engage with the substratum over which they grow. Instead, they use other molecules to build bridges across the cellular membrane. Key among such molecules are adhesion molecules that span the membrane from the inside to the outside. These adhesion molecules adhere to the substratum with their extracellular domains, and with their intracellular domains they make connections with actin-binding proteins that are connected to the actin cables. If the adhesion is good, the effect is to engage the actin cables of the filopodia to the substrate, which steers the whole growth cone in the direction of the engaged filopodia. The disassembly of actin filaments at the rear end of each filopodium breaks the connections between the substratum and the actin cables, while new connections are formed between the substrate and the growing front ends. Again, this is like a tank, where the treads at the rear disengage from the ground just as the treads at the front engage.

Imagine a growth cone with several filopodia spread out like the fingers of a wide-open hand. On the right side of the growth cone, the substratum is adhesive and good for gripping, while on the left side, the surface is a bit slippery. The filopodia on the right side that make contact with this sticky surface pull the entire growth cone right, while the filopodia on the left are having trouble getting a grip. The result is that the growth cone bears to the right. Consider another situation in which the substratum conditions are uniform, good for growth-cone progress in any direction, but this time, a molecular signal on the right side stimulates the growth cone to extend more filopodia on this side. The pull to the right becomes stronger as a result, and the growth cone turns right again. Similarly, a different molecular factor on the left may decrease filopodial extension, leading the growth cone to turn away—we call this "repulsion." Thus, a growth cone can propel itself forward, and it can steer itself.

The growth cones of every neuron are beginning to venture out on their various expeditions, ready to sense and respond to the molecular signs and cues that show the way ahead. And they intrepidly pull themselves through different terrains over long distances to destinations elsewhere in the brain.

Pioneers and Followers

Given the complexity of the brain wiring, Cajal reflected on how horrendously difficult it was going be to figure out how all these pathways are formed. He mused: "Since the full-grown forest turns out to be impenetrable . . . why not revert to the study of the young wood, in the nursery stage."[4] In this younger, simpler environment, pioneering axons navigate the earliest pathways in the brain, blazing trails that later axons will follow. As the brain matures, more and more axons join these initial pathways, which will become the information highways of the brain, the main axon tracts. From these "highways," some axons head off onto "roads," and some of the axons on the roads branch onto "lanes." The road map of the brain becomes increasingly more intricate as more axons enter the nexus and then branch off to create new routes. So, as Cajal suggested, some developmental neuroscientists began to look for the very earliest axons, the pioneers who blaze the trails.

The first observations of navigation by pioneering axons were made by Michael Bate at the Australian National University in 1976.[5] Bate found that at the tip of each developing leg in a locust embryo sits a pair of sensory neurons, and these two neurons send their axons up through the legs and into the central nervous system at a stage when there are no other axons around them. Surprisingly, these pioneering axons of the locust's leg nerve take a route that travels along a set of special cells that occur at regular intervals along the developing leg. They seem to act like a series of "stepping stones," as Bate called them. The last of these stepping-stone cells occurs just before the axons enter the central brain. At the University of California Berkeley, David Bentley noted that the distances between the stepping-stone cells are just small enough so that some of the longer filopodia from the growth cones of the pioneering axons can reach out to the next stepping stone cell while still clinging to the previous one. Bentley and colleagues were then able to show that these stepping-stone cells are truly critical for pathfinding, because when one was experimentally zapped with a focused laser microbeam, the growth cones of pioneer axons frequently stalled at the previous stepping stone, and sometimes they turned around and headed back out to the tip of the growing leg.[6]

Stepping-stone cells provide pathways for pioneer axons that enter the central nervous system from the leg, and the axons of these pioneer neurons become the pathways that later arising sensory neurons of the leg depend on as they head toward the central nervous system. Is this the kind of axon guidance that happens within the brain itself?

In the late 1970s, Corey Goodman, a PhD student of Bentley, went to the laboratory of Nick Spitzer at the University of California, San Diego. Spitzer, an expert neurophysiologist of the embryonic frog spinal cord, worked with Goodman to impale cells of the embryonic locust central nervous system with a microelectrode and fill the cells with a dye. When these preparations were put under a microscope, the entire anatomy of each filled neuron could be seen. Goodman and Spitzer teamed up with Michael Bate to show that in each segment of the ventral nerve cord, the neurons are uniquely identifiable because of the individual and specific patterns of its growing axons and dendrites. Many of the same types of neurons can be found from animal to animal and from segment to segment.[7] In his new lab at Stanford University, Goodman and colleagues began to look at the extending axons of particular neurons in the central nervous system to understand how individual axons navigate. For example, they saw that the growth cone of a neuron called the "G-neuron" first adheres to the axon of the C-neuron that recently pioneered a pathway crossing from one side of the nervous system to the other. When the growth cone of the G-neuron reaches the other side of the midline, it lets go of C and begins to probe the new environment, extending filopodia that touch several other axons going in different directions. The filopodia of the G-neuron's growth cone then grab onto one of these axons from a cell called "P" and use it to head in the right direction, toward the brain (figure 5.3). If this P axon is experimentally eliminated, the growth cone of the G-neuron patiently waits for the next P axon from the next segment to arrive. Meanwhile, the growth cone of the C-neuron grabs onto a different axon and heads posteriorly.[8]

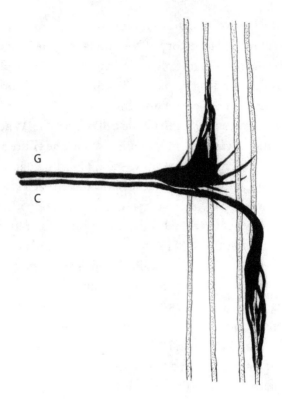

FIGURE 5.3. Labeled lines. C and G growth cones in a locust embryo choosing to grow in different directions along different axons.

The work of Bate, Bentley, and their colleagues hinted between pioneering growth cones and stepping-stone cells, and between different axons, that there are special kinds of adhesive interactions, which are critical for axon guidance. It became of great interest at that point to hunt down the molecular nature of the differences in cell adhesion in the developing nervous system and to see whether they were shared across a wide array of animals and potentially even humans.

Molecular Guidance

Goodman's laboratory began to search for the molecules that were responsible for differential adhesion between axons by raising antibodies to the membranes of the locust embryonic nervous system and then searching for antibodies that labeled small groups of tightly packed axons called fascicles. They then used the antibodies to purify the proteins. The first two such proteins they found in this way were called Fasciclin-1 and Fasciclin-2. Fasciclin-1 is found on some axon fascicles that cross the midline, while Fasciclin-2 is found on some axon fascicles that are running from head to tail. Follower axons, like the G-neurons mentioned above, can use these pathways. If they want to cross the midline, they express Fasciclin-1, and if they then wish to grow in a tail-to-head direction, they switch to Fasciclin-2. In this way, the different fasciclins correspond not to different axons but to different portions of the route.

Goodman visited Bate, who was then at Cambridge University, and together they found that it was possible to do the same sort of experiments on embryos of the fruit fly (*Drosophila*). Fruit fly mutants would provide access to the genes involved in axon guidance, and the genes would reveal the molecular nature of these guidance factors. *Drosophila* embryos are much smaller than locust embryos, so mapping the pioneers in *Drosophila* was difficult. Nevertheless, they managed to show that *Drosophila* has a very similar set of pioneering axons as the locust, only miniaturized. Goodman and colleagues then began to look for *Drosophila* mutants with embryonic wiring mistakes and soon began to find them. This was the start of tremendously exciting times for developmental neurobiology. We knew that soon the types of molecules that guide growing axons would be revealed!

Many of the mutants found in Goodman's lab turned out to affect genes that code for cell adhesion molecules (CAMs). Fasciclin-1 and Fasciclin-2 both turned out to be CAMs. One property that most CAMs share is that they are homophilic (they literally love themselves), which means they bind to the same CAM on the surface of another cell. In the developing embryo, if two axons express the same homophilic CAMs on their surfaces, they will tend to adhere to each other and make a bundle or fascicle of axons. The developing nervous system uses homophilic CAMs like a color-coding system: red axons stick to red axons, and yellow axons stick to yellow ones. Pioneer axons decorate themselves with specific sets of CAMs so axons that express the same CAMs can follow them. In this way, axonal tracts are built up in the central nervous system. In addition to following the axons that express the same CAMs, an early axon may also pioneer a new route during the last leg of its journey and add a new CAM to help future axons reach the same site. As more axons and more CAMs are added to the network, the simple scaffold of the first pioneers becomes a complex set of major and minor pathways through the brain. The idea emerged that axons travel to their destinations in a way that is like how Bostonians travel to the Zoo. They take the Green Line to Park Street, get off there, and switch onto the Red Line, getting off at Ashmont. From Ashmont, they take the number 22 bus to Franklin Park, and they walk the rest of the way themselves.

Local Guidance

The axons of retinal ganglion cells are true pioneers. They navigate from the retina to their targets in the brain without encountering any other of the early pioneering axons traveling in different directions. Even when an eye primordium is transplanted from an older to a younger frog embryo, such that retinal ganglion cells axons become the very first axons in the whole brain, they nevertheless navigate unerringly to their targets. These pioneer axons forge a pathway that will become the optic nerve and optic tract. The axons of the retinal ganglion cells cross the ventral midline at a region of the brain called the "optic chiasm" and then head dorsally near the border of the forebrain and midbrain. Most of them turn caudally to finally arrive at their targets in a region of the dorsal midbrain known as the optic tectum. Time-lapse movies of the growth cones of retinal ganglion cells in the brains of many model species of vertebrate animals show that they grow steadily at a relatively constant speed, pausing occasionally at decision points, but rarely straying off course.

How do these retinal ganglion-cell axons find their way? One possibility is that their target, the optic tectum, secretes diffusible "come hither" molecules. Then the growth cones, like bloodhounds, follow the scent to its source. This hypothesis seemed consistent with a set of experiments I once did. I transplanted eye primordia in amphibian embryos to different regions of the brain. No matter where they started out in the brain, the retinal ganglion-cell axons always seemed to orient their growth toward the optic tectum.[9] Soon after my study was published, Jeremy Taylor at Oxford University completely removed the optic tectum from a developing frog embryo before the axons of retinal ganglion cells had even grown out of the eye; yet when they did emerge, the retinal ganglion-cell axons still navigated beautifully all the way to the dorsal midbrain, only to find that the targets had been long gone.[10] So there was obviously another explanation for my results than long-range "come hither" molecules.

What this explanation was had been previously revealed in part by Emerson Hibbard at the California Institute of Technology in 1965. Hibbard focused on a set of "giant neurons" found in the hindbrains of fish and young tadpoles. These are the Mauthner neurons, named after Ludwig Mauthner, who in the mid-nineteenth century found that these neurons mediate a rapid escape response. Hibbard could easily identify their large axons on a microscope side. He saw that they crossed the mid-line and descended caudally down the spinal cord to innervate motor neurons along the opposite side of the body. When a Mauthner neuron is excited by a touch or vibration on one side of the body, the result is the almost immediate contraction of all the muscles of the other side, causing the animal to coil into a C shape, the starting position for a rapid dart of swimming. Hibbard's conceptually simple experiment was to take a little slice of the neural plate that was fated to become hindbrain tissue in a salamander embryo, rotate that piece 180 degrees tail-to-head (caudal-to-rostral), and then transplant it into another embryo of the same age. In these experiments, the host embryo developed an extra piece of hindbrain that was rotated, and in this transplant tissue, there was an extra pair of Mauthner neurons. When these extra Mauthner neurons sent out their axons within the rotated bit of hindbrain, they first travelled in the wrong direction, up toward the midbrain rather than down toward the spinal cord. However, once they exited the rotated piece and found themselves in unrotated tissue, they made dramatic U-turns and descended (figure 5.4). Hibbard surmised that the orientation of axon growth must be influenced by interactions with the local environment.[11] After reading about Hibbard's early work, I then did a similar experiment on retinal ganglion-cell axons, which showed that when these axons entered a piece of clockwise-rotated tissue on their way to the optic tectum, they also turned clockwise as though they were paying attention to local cues in the rotated piece. But when they exited the rotated piece and "discovered" that they were

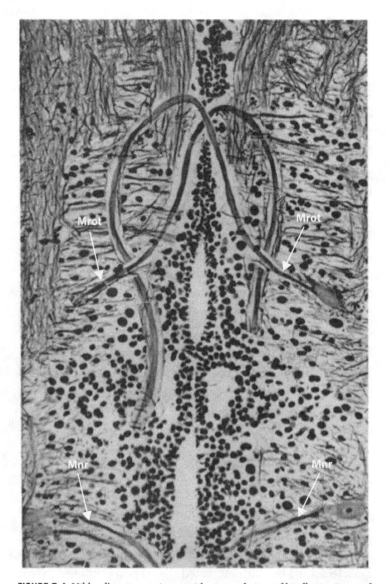

FIGURE 5.4. Hibbard's 1965 experiment with a rotated piece of hindbrain. Axons of reversed-oriented giant Mauthner cells (Mrot) cross the midline, head upward toward the brain, and then once they exit the rotated tissue, they turn and head tailward toward the spinal cord. Axons of the nonrotated Mauthner cells (Mnr), seen near the bottom of the micrograph, cross the midline and head tailward as normal. Reprinted from E. Hibbard. 1965. "Orientation and Directed Growth of Mauthner's Cell Axons from Duplicated Vestibular Nerve." *Exp Neurol* 13: 289–301.

not where they expected to be, they made corrective turns and found their way to the optic tectum.[12] Such cut-and-paste experiments suggest that pioneer axons navigate by reading localized cues rather than sniffing out distant signals.

The process of axonal wiring up of the nervous system could be compared to the stepwise growth of a hamlet as it transforms into a giant metropolis. First to arrive are the pioneers who blaze their own trails using compasses and their knowledge of geography. They read the local topography and understand the lay of the land: how the ravines and rivers run to the lake. The pioneers are followed by others. The pathways that the pioneers first walked become well-trodden, and the bigger ones become roads and then streets—Yonge St., which runs North–South; and Bloor St., which runs East–West. As the town grows into a city, the streets lengthen and widen into highways, many new streets are made, subway lines are formed, and there is a hockey team. And if you have a map of the city, the means to pay for a public transport token, and you read the signs, you can get from almost anywhere to a Leafs' game.

Attraction and Repulsion

In the 1980s, Andrew Lumsden and Alun Davies at University College London were asking how certain sensory axons find their way to the base of the whiskers in a mouse. Mice build up a touch-based map of their local world by whisking their whiskers, which are heavily innervated. The region of epidermis that contains the base of the whiskers, called the "maxillary pad," is the target of many sensory axons of the fifth cranial nerve. Lumsden and Davies put these sensory neurons and the maxillary pad close to each other in a tissue culture dish at a stage preceding their contact. The axons grew directly toward the maxillary pad, and they were not fooled or diverted toward other tissues, which Lumsden and Davies sometimes added to the co-cultures. The researchers inferred that the maxillary pad releases diffusible molecules that attract these sensory axons.[13] Chemoattraction is a process in biology whereby a cell detects a local chemical gradient of a substance and then moves toward it. For example, chemoattraction helps white blood cells move to sites of infection. Because it came from the maxillary pad, Lumsden and Davies nicknamed the unknown chemoattractant "Max Factor."

The existence of such a chemoattractant might seem to contradict the work described in the previous section, which suggested that targets do not attract incoming axons from a long distance. But because attractants in the brain tend to stick to the substratum of extracellular material soon after they are secreted, they are detected only when axons are near the sites of secretion. In addition to attractants in the local environment, cells can also secrete chemorepellents, which growth cones avoid if they get too close. Thus, while the maxillary pad is secreting chemoattractants, nearby tissues secrete chemorepellents, so there really is only one good choice for these sensory axons, which is to enter the maxillary pad and innervate the whiskers.

The 1990s heralded a new era of rapid discovery of the molecular nature of life, which followed from advances in modern molecular genetics and genetic engineering. Scientists were finding and sequencing the proteins and genes responsible for almost any process that could be tested in a culture dish or model organism, such as nematodes, flies, and mice. This work enabled searches for the molecular nature of chemoattractants and chemorepellents. In 1990, Edward Hedgecock and colleagues at the Johns Hopkins University found three genes that affected the navigation of pioneer axons in the tiny nematode *C. elegans*.

These genes came from a set of mutants with uncoordinated locomotion, so they all were called simply unc (*unc*oordinated) mutants. One of these three unc mutants affected the navigation of axons traveling dorsally, another affected the navigation of axons traveling ventrally, and yet another affected both dorsal and ventral axon navigation. Hedgecock and colleagues then cloned these genes. The one unc gene that affected both dorsal and ventral navigation turned out to encode a novel secreted protein that guides axons.[14] Marc Tessier-Lavigne and colleagues at the University of California, San Francisco, soon found vertebrate homologues of this unc gene and dubbed the guidance factor that the gene makes "Netrin" (after "*Netr*," the Sanskrit word for "one who guides").[15] The other two unc genes made receptors for Netrin. Growth cones that had one of these receptors were attracted to Netrin, while growth cones that had the other were repelled by Netrin.

Netrin is a member of a family of well-conserved and evolutionarily related proteins that are involved not only in axon guidance but also in the migration of neurons and cells in many other body tissues, including various ducts and blood vessels. Mutations in Netrin genes therefore disrupt the morphological development of many tissues and cause a variety of syndromes in humans. Mutations in the first Netrin gene have rarely been seen in humans possibly because the Netrin gene is essential for development. However, some mutations disable Netrin function only slightly and are not lethal, but they are associated with abnormal axon crossing in the spinal cord and a corresponding behavioral disorder characterized by involuntary movements of one hand that mirrors the intentional movements of the opposite hand.[16]

At the same time as the Netrins were being discovered, Jonathan Raper and colleagues working at the University of

Pennsylvania discovered a repulsive guidance factor. They originally called this factor "Collapsin," because when active growth
cones are exposed to tiny amounts of Collapsin, it causes them
to withdraw all their filopodia and collapse into a simple bulb
shape.[17] The growth cones lose their grip on the surface, and
the axon quickly slips backward. When the smallest amount of
this chemorepellent is applied to just one side of the growth
cone, it loses its filopodia on that side and turns away, as if re-
pulsed. Collapsin became the first identified member of yet
another large family of axon guidance factors that was also
found by Goodman and colleagues in *Drosophila* and called the
"Semaphorins" (after semaphore—a system of signaling by
flags or arm positions used to convey information at a distance).
There are 20 different Semaphorin genes in the human genome,
and each gene makes a slightly different version of one of these
Semaphorins. And as is the case for the Netrins, there are also
many different receptors for these Semaphorins. One can imagine that the growth cones of different axons respond to the
Netrins or Semaphorins they see in their own ways: some are
attracted, some are repelled, and some indifferent. And from
that perspective, one can begin to envisage how a huge diversity
of choices can be made by navigating growth cones heading in
different directions.

Since the identification of the Netrins and Semaphorins,
many more axon guidance factors have been discovered
in many laboratories. If one looks at the early nervous system
and makes a map of where all these guidance factors are made,
one sees that the embryonic nervous system is virtually covered
in guidance factors of various sorts. And like different people
might use the same map of the city to go to different places,
different axons may use the same molecular patterns in the
brain to grow to different targets. So, the early brain is like a

three-dimensional patchwork quilt of cues for pioneer axons, which may encounter a new molecularly distinct region about every 20–50 microns. As not everyone is going to the hockey game, the important thing is to know where you want to go and where you are now on the map. Growth cones are sophisticated readers of the molecular maps in the developing brain, continuously integrating combinations of attractants, repellents, cell adhesion molecules, and other potential guidance cues when making navigational decisions. It is, after all, their one job in life.

Intermediate Targets

The mid-nineteenth century saw wagon trains of weary travelers moving West across America, over the Rocky Mountains to Oregon and California. Most of these pioneers of the West first headed toward a huge granite hill called "Independence Rock" in the Sweetwater Valley of Wyoming. Independence Rock marked the approximate midway point for those on the Oregon Trail. Here, the pioneers often stopped for a while to rest and get prepared for the next stage of their journey. They often scratched their names into the rock along with the dates they arrived. But they could not tarry long, especially if they arrived there at the beginning of July, or they might not make it over the Rocky Mountains before the snows came and made the mountain trails impassable. Growth cones have similar issues. They may also be attracted to a place en route, but they must not stop there for long. This is merely an intermediate target. Axons must leave such attractive intermediate targets and move on to take on the next leg of their journeys.

The issue of how axons first go to and then leave an attractive intermediate target has been extensively investigated at the ventral midline of the nervous system, which serves as an intermediate target for many axons that cross from one side of the brain to the other. Neuroanatomists call such axon crossings of the midline "commissures." Commissures are a common feature of our nervous system because of the need to coordinate sensory and motor functions across the two sides of the body. The ventral commissures of the vertebrate spinal cord are pioneered by neurons that first send their axons ventrally. Near the ventral midline, they encounter attractive factors, such as Netrin. They then cross the midline and usually grow either up toward the brain or down toward the tail on the other side of the spinal cord, and they never cross the midline again. That is, once on the other side, the midline attractants are no longer of any interest to the growth cones. How can that be?

The first insight into this problem came from a mutant fruit fly found by the Goodman lab in their search for axon wiring mutants in *Drosophila* embryos. In this mutant, called "Roundabout" (Robo), axons cross the midline over and over again and go around in circles, as I sometimes do on British roundabouts. The Robo gene codes for a receptor that senses a chemorepellent called "Slit" that is expressed at the ventral midline. Before crossing the midline, the growth cones of commissural neurons make receptors only for midline attractants but not for repellents. But once they cross the midline, they begin to make Robo, the receptor for Slit, and the midline becomes more repulsive than it is attractive. In Robo mutants, these axons cannot sense the chemorepellent, so they cross and recross the midline. Whether we consider a fly or a human nervous system, the concepts and many of the guidance molecules are similar: the commissural axons are initially attracted to the midline, but as they cross, they change, so that the midline becomes unattractive, or even repulsive.[18] The ventral midline is just one of many intermediate targets in the nervous system that pioneering axons must grow to and then leave. This strategy breaks their long journeys into manageable segments, moving from one intermediate target to the next without much stalling and without turning back.

In 1987, my longtime frequent collaborator (and amazing wife) Christine Holt and I were on sabbatical in the laboratory of Friedrich Bonhoeffer at the Max Planck Institute for Developmental Biology in Tübingen, Germany. We were making movies of the growth cones of retinal ganglion cells navigating to their targets in the dorsal midbrain in frog embryos. To do this, Christine would carefully make a tiny slit above one eye bud and insert a needle carrying a little fluorescent dye, which she would transfer to a few cells. The next step was to flip the embryo over onto a microscope slide, so we could make high magnification time-lapse recordings of the growth cones of retinal ganglion-cell axons that had crossed the midline at the optic chiasm and were on their way to the optic tectum or the dorsal midbrain. A successful recording session would continue through the day and into the night. One day, when Christine was flipping over a successfully labeled embryo, her hand jerked a bit, and she accidently ripped off the tiny eye bud that had the labeled retinal ganglion cells. By the time this happened, though, the first retinal ganglion-cell axon had already left the eye, crossed the ventral midline at the optic chiasm, and was just now in the act of crawling dorsally up the other side of the brain. At this stage, the specimen would have been perfect for recording, except that we had accidently detached the growth cone from its cell body. We debated for a while about whether to waste the day looking at this poor growth cone. We were thinking that such axons would surely just die, and I was for starting again, but Bonhoeffer, whose guests we were, encouraged us to look at it anyway, to see what would happen. The time-lapse movie we made hugely surprised us all. The detached growth cone continued to grow and to navigate correctly for several hours, and other growth cones, whose axons we later deliberately severed, revealed that these growth cones are remarkably autonomous when separated from their parent cell body and nucleus.[19]

Growth cones, it turns out, have all the machinery needed for making new proteins and degrading old ones. They continuously synthesize new proteins using the thousands of different messenger RNA molecules that have traveled from their origin in the nucleus to these far outposts of the neuron. Many guidance cues, both attractive and repulsive, stimulate protein synthesis in the growth cone within minutes of binding their receptors. Holt and colleagues found that this new synthesis of proteins is crucial to the responses that a growth cone makes to guidance cues and to what it does next.[20] When a growth cone arrives at an intermediate target, it can rapidly reset its navigational priorities, such that the present target becomes less attractive, while the next target becomes more attractive.

Generally, guidance cues such as Netrin, when attractive, lead to an increase in the local synthesis of proteins involved in adhesion and the assembly of the growth-cone cytoskeleton. When a growth cone senses an attractive guidance factor on one side, the local protein synthesis on this side causes enhanced growth in that direction, and the growth cone turns toward the attractant. In a repellent mode, the same guidance factors lead to a decrease in the synthesis of these proteins and an increase in proteins that favor disassembly of cytoskeleton. So, when a growth cone senses a repulsive guidance factor on one side, it decreases growth in that direction and turns away. As neat and simple as this logic might at first seem, it also points to the challenge that faces future workers in this field. If growth cones are autonomous machines capable of continually rede-fining themselves as they navigate through an embryonic brain, which is packed full of guidance signals and adhesion molecules of differing attractiveness, how are we ever going to figure it all out?

Regeneration

In 1928, Cajal wrote, "Once the development was ended, the founts of growth of the axons and dendrites dried up irrevocably. In adult centers, the nerve paths are something fixed, ended, im-mutable. Everything may die, nothing may be regenerated. It is for the science of the future to change, if possible, this harsh decree."[21] When Marc Buoniconti was a young college football star in 1985, his spinal cord was damaged in a game. Since that moment, he has been unable to move any muscle below his neck. Inspired by the prospect of finding a cure for Marc's injury as well as that of many others who have suffered such debilitat-ing spinal cord traumas, Marc's NFL Hall of Fame father, Nick Buoniconti, joined Barth A. Green to establish the Miami Proj-ect to Cure Paralysis.[22] The Miami Project is one of several major research and treatment centers around the world that are fo-cused on this single mission. Yet sadly, in adult humans, the prognosis for recovery from serious spinal cord damage or se-vere traumatic brain injury is still poor. This is not an indictment of the science, because repairing a broken nervous system is probably one of the biggest challenges in all of medicine. Re-search on laboratory animals and neurons in tissue culture has revealed that adult mammalian neurons are intrinsically less able to regrow than are young neurons. They seem to have lost the mojo of their youth. Research has also shown that the site that bears the scar of the neural damage is filled with non-neuronal cells and extracellular materials that have inhibitory effects on regrowing axons. Finally, many of the guidance cues that are used by neurons to find their proper targets during their initial outgrowth in the embryo are often not available to guide regrowing axons in the adult.

Scientists have made serious progress in understanding some of the details underlying these challenges for regeneration, but this has not yet led to much in the way of full cures for paralysis. Nevertheless, it is still the hope of people working in this field that the investigation of the mechanisms of axon growth and guidance during development will be useful for finding ways to help severed axons regrow in adults, so that one day we may be able to cure nervous system injuries like Marc's.

Wiring up is one of the great feats of the developing brain and continues to be one of the great challenges of neuroscience and medicine. Accurate axonal navigation by the first pioneers and their billions of followers is responsible for ensuring that the right information gets to the right places for processing and computation. The guidance cues that growth cones use to navigate may be in the form of membrane-bound adhesion molecules, or local chemoattractants and chemorepellents. Whether they are trailblazing new routes or using other axons as a kind of public transport system, they frequently change and update their priorities as to where to head next. In the end, almost all of them reach their destinations. It is here that they will find their synaptic partners and make connections, allowing the brain to begin to fire up.

6

Firing Up

In which we witness the momentous moment in the lives of two neurons when an axon of the first meets a dendrite of the second and the two recognize that they are meant for each other. They stick tightly together and seal the deal with a synaptic kiss.

Specificity

A baby's first kick inside the womb, at about halfway through gestation, is a memorable moment in a pregnancy, a new source of connection with the unborn child, a potential topic of conversation that may prove embarrassing to that child a few years later. When a baby starts kicking, what does this mean in terms of nervous system development? One thing it certainly means is that the axons of motor neurons have reached the end of their travels and have started to make synapses with the muscles in the leg. It is only once these synaptic contacts are made that the kicking can begin. Like the axons of leg motor neurons, billions of other axons in the brain have also come to the end of their journeys and have started to make synapses with one another. Having arrived at their specific destinations, the growing axons shed their growth cones and begin to sprout branches that weave their way through the thousands or millions of target neurons that surround them. The target neurons have been waiting for axons to arrive, reaching out their dendrites for them. The branching axon terminals are searching among this forest of growing dendrites for their most suitable synaptic partners and vice versa.

The connections in the nervous system are made with incredible precision, yet nothing was known about how this specificity is achieved until the 1920s. Then, at the Academy of Sciences in Vienna, a young developmental biologist named Paul Weiss started working on this challenging problem. As a graduate student, Weiss had been studying the regeneration of nerves in newts and noted that when a leg nerve regenerates, the animal regains its ability to move the leg in a purposeful and coordinated way, with all its reflexes intact, and in perfect rhythm with the other limbs. To explore whether sensory feedback has a role in reestablishing this coordination, Weiss allowed only the motor axons, but not the sensory axons, to regenerate, so that the animal could not feel the regenerated limb. Yet once again, coordinated movement was perfectly restored. So sensory feedback was not required. Weiss proposed an explanation that he called "myotypic specification." According to myotypic specification, motor neurons are not initially very choosy; they innervate limb muscles rather randomly—any motor neuron can innervate any muscle. Cross-innervation experiments had already shown this to be so. The next step involves communication between the muscle and the motor neuron. The muscle cell communicates something like this: "Hi, you have just made a synapse with me, gluteus maximus." The motor neuron then uses this information to wire up its connections in the spinal cord properly. Myotypic specification was particularly useful for explaining the results of

nerve-crossing experiments. For example, Weiss and his colleagues would cut a nerve that innervated an extensor muscle in
a newt leg, and then attach it to a nearby flexor muscle. When
such cross-innervated newts recovered movement of the leg, the
movements were immediately abnormal, reversed in a way,
proving that the cross-innervation was successful. But after a
week or so, the cross-innervated legs began to resynchronize and
regain all their previous coordination.

Weiss's concept of myotypic specification was beautiful in a
way, because it could be generalized to help understand wiring
up the brain as a whole. If specificity could reside in the muscles, might it not also reside in other cells and be transferred to
any neurons that synapse with it? A developing synapse would
be a place to exchange information about cell identities. In this
expanded version of Weiss's concept, known as the "resonance
hypothesis," each neuron tells the neurons that innervate it
what it has successfully done so far, and these neurons then pass
that information to the neurons that innervate them. "Hello,
yes, I am an inhibitory neuron Type X-24-sigma, and I have
connected to a motor neuron that has just told me that it is connected to the gluteus maximus." And so on.

Unfortunately for the resonance hypothesis, soon after Weiss
moved to the University of Chicago, he took on a graduate student by the name of Roger Sperry, whose work would undermine his own.[1] In his PhD work in Weiss's lab, Sperry used
young rats to see if myotypic specification applied to mammals.
At early stages of postnatal development, rats can regenerate
their peripheral nerves to some extent. Sperry cross-innervated
flexors and extensors in their legs. What he found was that unlike in newts, there was no trace of recovery of normal movement over the course of a year or more in these rats. The experimental rats always moved their cross-wired limbs oddly, and
when these animals were put into an enclosure containing a

small, electrified grid, the cross-wired limb would press harder on the grid, instead of lifting off as quickly as possible, which is what the three other legs did. This aberrant reflex response did not renormalize at all over the course of more than a year.

That this result is so different than the results of nerve-crossing experiments in newts made Sperry wonder whether there was another explanation for the results in newts. What if in the experimentally cross-wired newts, the crossed axons were somehow uncrossing themselves? What if they were navigating their way back to their true homes (i.e., their original muscles)? To test this idea, Sperry began to do cross-innervation experiments in a variety of animals (fish, newts, and frogs) while taking great pains to prevent any possible re-innervation from the original nerves. In these cases, there was no recovery of normal movement. Sperry's explanation for the recovery of normal coordination in the experiments done by Weiss was that the original nerve found its way back. Indeed, it is now known that in such situations in amphibians, the original nerve does grow back to its own muscle and kicks out any foreign synapses. This work showed that the idea of myotypic specification of regenerating motor neurons was probably wrong, due to a simple error of interpretation of what was really happening.

When I was a graduate student in the 1970s, I was fortunate enough to be a teaching assistant in an introductory neurobiology course that Sperry taught at the California Institute of Technology. I remember one of the students asking Sperry whether cross-innervation procedures had ever been done in humans. Sperry explained that in cases of facial nerve damage, nerves can be rerouted during surgery to help the patient regain some control and muscle tone. Humans can even learn to command the "wrong" muscle to do the right thing with conscious effort (e.g., learn how to smile), but natural expressions of emotions and reflexes that arise unconsciously often cause the wrong muscles to contract in such patients, leading to inappropriate movements and expressions.

Though Weiss and his followers had certainly misinterpreted the cross-wiring experiments, there remained one slim hope for the concept of myotypic specification. Perhaps the specification happened when the connections were originally made. These first connections could still be made rather indiscriminately with target muscles, which would then tell the connections who they were. Then, if at some later stage, an experimenter cut the nerve and allowed to it regenerate, the nerve would by then already "know" who it was and which muscle to grow back to. However, it was not easy to test this idea. But then, in the 1980s, Lynn Landmesser at Yale University found a way to label small groups of neurons in the chick embryo even before they started to send out axons. In a key experiment, Landmesser and a research colleague, Cynthia Lance-Jones, rotated pieces of the spinal cord in these embryos at a stage of development when the axons of motor neurons had not yet begun to grow out. They saw that the axons of these virgin motor neurons grew to their appropriate virgin muscles, even if they had to take unusual routes to get there. These motor neurons were by no means indiscriminate. They intrinsically "knew" which muscle to connect with from the start. Landmesser and Lance-Jones concluded: "Motor neurons possess specific identities prior to axon outgrowth."[2] This was the final nail in the coffin of myotypic specification.

Chemoaffinity

Having helped to undermine the prevailing hypothesis for the wiring up of the nervous system, Sperry began to search for new insights into this process. He undertook a series of experiments, the results of which would lead to a new theory for the specificity of neural connections. These experiments involved the sense of vision and the connections between the eye and the brain. The first experiments were simple. He cut the optic nerve of a newt and waited for it to regrow to its main central target, the optic tectum of the dorsal midbrain. When vision returned, the animal saw normally. It would snap upward when a lure was held above it and would snap downward when it was held below. All good! This was much like a limb regaining perfect coordination when the nerve regrew. The next thing Sperry did was analogous to the cross-innervation experiments that he had done with motor nerves. In this series of experiments, he loosened an eyeball of a newt in its socket, rotated it 180°, and then sewed it back in, upside down. The question he was asking with this experiment was whether vision would be rotated for the animal. And if vision was somehow rotated at first, could the nervous system adjust so that the newt would eventually snap in the right direction to catch the worm? The experiment yielded the clearest of results on both counts. The answer to the first question was "yes." When vision returned, the newts behaved as if their world was back-to-front and upside down. Sperry did the same kind of experiment with numerous species of newts and frogs with similar results. He then cut the optic nerve before rotating the eye, so it might have a better chance to find more functionally appropriate connections when it regenerated. But the results were always the same, as Sperry described it:

When a fly was held in front of the animals within easy jumping distance, they wheeled rapidly to the rear instead of striking forward. Contrariwise when the lure was held in back of them and a little to the side they struck forward into space. When the animals came to rest in such a position that the lure could be presented well below eye level, they tilted the

head upward and snapped at the air above. When the lure
was held above the head and a little caudad to the eye the
animals struck downward in front of them and got a mouth-
ful of mud and moss.[3]

The answer to the second question was equally clear. The answer
was "no," the operated animals never recovered. They snapped
in the wrong direction throughout the remainder of their lives.

A point-to-point or topographic projection from the retina
to the optic tectum arises because axons originating from
neighboring positions in the retina make synaptic connections
with neurons on neighboring positions in the optic tectum.
This topographic pattern of neural connectivity preserves the
continuity of visual space in the physical architecture of the
brain. The orderly mapping of the retina onto the optic tectum
must be achieved through the ability of retinal ganglion-cell
axons to make synapses at precise topographic positions along
the two axes of the optic tectum: anterior-to-posterior and
medial-to-lateral. Sperry therefore boldly postulated the existence of matching molecular gradients across the retina and
across the optic tectum. For example, a gradient might exist of
one molecule in the retina that binds another molecule that
forms a gradient in the optic tectum. One gradient might be
composed of a ligand, the other of a receptor for this ligand.
This process could create an orderly map along one axis. To
cover the whole of visual space, Sperry therefore proposed the
existence of two such gradients "that spread across and through
each other with their axes roughly perpendicular." These gradients would stamp each neuron of the optic tectum with its appropriate latitude and longitude in a kind of chemical code.
Retinal ganglion cells from particular coordinates on the retinal
surface would recognize partner cells at coordinates in the optic

tectum that had matching chemical values. This was Sperry's "chemoaffinity" hypothesis.[4]

Now that Sperry's concept of chemoaffinity stood in place of Weiss's concept of resonance, it also had to stand up to experimental challenges. One of the first things that needed to be checked out about the new hypothesis was whether retinal ganglion cells, like motor neurons, were specified from birth or gained their specific identities through their original interactions with their partners in the optic tectum. Sperry's experiments were all done on regenerating axons, which meant that it was possible that the axons of the retinal ganglion cells were just following their old routes. In the early 1980s, eye bud rotation experiments, in which the future eye was rotated long before axons ever emerged, showed that the adult animals that developed from the experimental embryos also see the world upside down and backward through this eye. Just like the case for motor neurons, it appears that retinal ganglion cells are also specified to find their appropriate synaptic partners long before they begin to send out axons.

Chemoaffinity had stood up to this first experimental challenge, but there was another reasonable explanation for how the retina connected topographically with the optic tectum that did not need to invoke the idea of chemoaffinity. This idea was based on the order in which axons enter the target area. Imagine the optic tectum as a concert hall filling up with avid fans. Seats are unassigned, but ushers direct the first arrivals to the front row seats, the next to enter go to the next row, and so forth. The axons of retinal ganglion cells do arrive in the tectum in an orderly fashion, from dorsal to ventral. So, timing of arrival rather than chemoaffinity could in theory establish the initial topographic connectivity. However, experiments in which the order of arrival was altered failed to disturb the normal topographic mapping.[5] Chemoaffinity had withstood another challenge. As more and more experiments ruled out other alternatives to the hypothesis, molecular biologists began to take chemoaffinity seriously enough to start searching for the responsible molecules.

Gradients of Eph and Ephrins

Though many laboratories used various strategies to find the chemoaffinity molecules, it was not until 1987, more than 35 years after chemoaffinity was first proposed by Sperry, that Friedrich Bonhoeffer and his colleagues at the Max Planck Institute in Tübingen, Germany, made a significant breakthrough. They developed a tissue culture method whereby retinal ganglion-cell axons were presented with a choice to grow on membranes from either one or the other of two pieces of the optic tectum. To do this, they removed the optic tectum of a chick embryo and cut it into three pieces: anterior, middle, and posterior thirds. They isolated the membranes from these pieces and used a microfluidic device that controls streams of fluids just microns wide to make tiny striped carpets from these membranes. Finally, retinal ganglion-cell axons from different regions of the eye were placed on these micro-striped carpets of membranes, and Bonhoeffer and colleagues noted the choices that the axons of the retinal ganglion cells made (figure 6.1). The most obvious choice was made by axons that originated from the temporal part of the retina (the part farthest from the nose). Temporal retinal ganglion cells normally send their axons to the anterior part of the optic tectum, and on the striped carpets of membranes, they much preferred to grow on membranes from the anterior tectum rather than on those from the posterior tectum. The temptation was to think that the

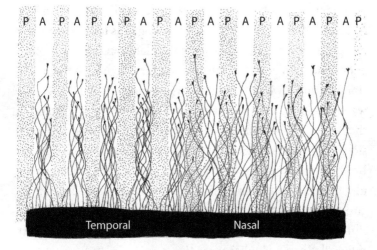

P A P A P A P A P A P A P A P A P A P A P A P

Temporal | Nasal

FIGURE 6.1. Bonhoeffer's 1987 striped carpet experiment. Axons from the temporal retinal avoid posterior (P) membranes of optic tectum, whereas axons from the nasal retina grow equally well on both posterior and anterior (A) membranes.

membranes of the anterior tectum were more attractive to these axons than were posterior membranes. However, Bonhoeffer and colleagues found that the temporal axons of the retina were not particularly attracted to anterior membranes. Instead, they were avoiding certain proteins that were present in posterior membranes. When these repulsive proteins were removed, the axons grew equally well on both anterior and posterior membranes. Bonhoeffer and his team then turned to molecular biochemistry to identify the repulsive molecule, and found that it formed a smooth gradient in the optic tectum strongest at the posterior pole of the tectum.[6]

While Bonhoeffer and colleagues were purifying this repulsive molecule, John Flanagan's lab at Harvard University was searching for molecules that bound to a large family of "orphan" receptors known as Ephs. An orphan receptor is one whose ligand remains unknown. Flanagan had developed a clever molecular strategy to find the mysterious ligands for these orphan receptors. In one set of experiments, Flanagan noticed a gradient of one such ligand in the optic tectum, highest at the posterior pole and lowest at the anterior pole; moreover, he saw there was a matching gradient of the Eph receptor for this ligand in the retina, highest on the temporal side, lowest on the nasal side.[7] It turned out that the repulsive molecule identified by the Bonhoeffer lab and the ligand of the Eph receptor that Flanagan identified were the same. The ligand was christened "Ephrin," and the gradient seen in the retina was now the unorphaned Eph receptor for Ephrin. The first of Sperry's chemoaffinity molecules had been discovered!

Sperry had hypothesized that two roughly perpendicular gradients of chemoaffinity in the tectum would be essential to wire up the retina to the optic tectum in a way that preserves a two-dimensional map of the visual world. The Ephrin that Bonhoeffer's and Flanagan's labs had identified worked along just one axis: anterior-to-posterior. If Sperry was correct, there should also be a second gradient of chemoaffinity along the perpendicular axis. Indeed, it was not long after the discovery of Ephrin (now called "Ephrin-A1") that other Ephrins were found, and they were matched up with different Eph receptors. One of these pairs, an Ephrin-B and its receptor, turned out to provide the perpendicular gradient along the medial-to-lateral axis of the optic tectum. This molecular topography, based on perpendicular gradients of Ephrins and Ephs that guide retinal axons to their appropriate topographic position in the optic tectum, is in striking agreement with Sperry's chemoaffinity hypothesis as he originally formulated it decades earlier. Indeed, different Ephrins and Eph receptors are now known to be involved in setting up orderly connectivity patterns between many other regions of the nervous system. For example, Ephrins are largely responsible for topographic patterns of connectivity that create maps of the body surface in the somatosensory pathways of the brain. Regions of the developing brain paint themselves with patterns of these ligands and their receptors in reciprocally graded arrangements that foretell patterns of synaptic connectivity between these regions.

In contrast to the retina, the sensory neurons in the ear are organized according to sound frequency. The cochlea is the auditory equivalent of the retina in that it is the first station of the nervous system to pick up sounds, and it is organized in a topography of pitch: high pitch at the base of the cochlea and low pitch at the apex. Ephrins and Ephs help preserve this topography of pitch in some brain regions. Other regions in the hindbrain deal with other aspects of sound. Some compute the difference in the time of arrival of a sound between the two ears. If a sound arrives a little earlier in your right ear than in your left, you look to your right to find the source of the sound. The auditory regions of the hindbrain that compute these time differences have the information to construct a map of auditory space from right to left. Other regions in the hindbrain compute whether the sound is coming from above or below. These neurons then send their outputs to the midbrain, where a map of auditory space is constructed. Again, Ephrins and their Eph receptors are involved in constructing these innate maps of space in the brain.

Cell Adhesion

Topographic maps are common in the connectivity patterns between regions of the brain, and gradients of Ephrins and Ephs are a good way to ensure appropriate initial mapping. But other regions of the nervous system use different strategies. For example, muscles are not organized in any simple topographic way in the larval stage of the fruit fly, when it is just a little white maggot eating its way through a banana. Each segment of the animal has 30 different muscles on each side of its body, and each segment of the body is innervated by a segmental nerve containing the axons of 30 motor neurons. As the muscles criss-cross one another at various angles, there is no obvious mapping of neuron position to muscle attachment that could be easily assigned through the smooth gradient of a molecule like Ephrin. In the absence of such a topography, how do the motor neurons find the right muscles? It seems like a simple problem of 30 by 30 matching, but 30 neurons can theoretically link up to 30 muscles in an astronomical number (30 factorial—more than 10^{32}) of ways, each of which would ensure that every motor neuron ends up with just one muscle and every muscle gets its own motor neuron. Yet only one of these ways is correct. How then does this system attain the perfect matching between the right motor neuron and the right muscle? The answer to this combinatorial problem seems to be a combinatorial solution, based on different combinations of various guidance and recognition molecules in different amounts. The guidance factors may be repellent or attractive and are key to bringing the axons of the motor neurons to the vicinity of their intended targets. The recognition of axon and target is certified through combinations of cell adhesion molecules (CAMs) of the same type that were described as having a role in axon guidance (see chapter 5). A motor neuron and a muscle cell with the right combinations and guidance cues and a few such homophilic CAMs can be matched.

The power of using matching combinations of homophilic CAMs as a driver of synaptic specificity is amazing.[8] Consider the simple knee jerk reflex. It is the reflex that causes your foot to kick when a doctor taps the tendon just below the kneecap with a small rubber hammer. This transiently stretches the quadriceps muscles of the thigh. In response, the quadriceps muscles contract, leading to the small reflexive kicking movement. Consider how this reflex develops. Each quadricep muscle has a set of motor neurons in the spinal cord that innervate it. Each muscle also has its own set of stretch receptor sensory neurons that send their axons into the spinal cord. The stretch receptor neurons and the motor neurons of each muscle express similar combinations of homophilic CAMs, enabling the sensory axons of any muscle to find the dendrites of the motor neurons to that muscle among the dense throng of axon terminals and dendrites in the developing spinal cord. The synapses are made with such precision that when stretch-sensitive neurons send signals along their axons into the spinal cord, they activate the motor neurons that innervate that same muscle. If a muscle is stretched a little, the receptors respond a little, and they activate the motor neurons just enough to contract the muscle to bring it back to its intended length. This simple reflex circuit allows us to stand up with our eyes closed and to automatically readjust our postures when we are pushed or given something heavy to hold.

The specificity in synaptic connectivity goes even deeper than neurons hooking up with their intended synaptic partners because synapses are organized on a subcellular level that takes account of the functional anatomy of the neurons. For example, the distal parts of dendrites (those parts that are farthest from the neuron's cell body) are preferentially innervated by excitatory axons, whereas the proximal parts of the dendrites (those closer to the cell body) tend to be preferentially innervated by inhibitory axons. Inhibition at the base of a dendrite is a good arrangement for vetoing excitatory signals originating in the distal parts of the dendrite from reaching the cell body. One type of inhibitory neuron takes advantage of a neuron's basic anatomy in a dramatic way by making its synapses right on the very beginning of the axon itself. The axons in question are those of the giant neurons of the cerebellum known as Purkinje cells and the local inhibitory neurons known as basket cells. The initial segment of the axon is the most effective spot for blocking output from a neuron. The Purkinje cells make a particular CAM that matches one that the basket cells make. The Purkinje then concentrates this CAM so that the initial bit of its axon has the highest amount. The terminals of the basket cells use this CAM to slide into position on the Purkinje cell as they begin making synapses just there.[9] The logic of building specificity through a system that relies on mutual adhesion through the expression of combinations of homophilic CAMs has the potential to allow axon terminals to create an incredibly selective set of synaptic connections between optimally matched partners, and even between parts of those partners.

Making Synapses

A mature synapse is a functional structure composed of three main cellular components: the presynaptic part from the axon terminal; the postsynaptic part from the target cell; and the glial cell, which often wraps part of itself around the synapse. Most of the work in our brains is done through chemical synapses, which may be excitatory or inhibitory. In a chemical synapse, the presynaptic element is packed with tiny spherical vesicles. Each of these tiny vesicles contains thousands of neurotransmitter molecules. The postsynaptic element of the synapse exposes a membrane packed full of receptors for these neurotransmitter molecules. When a neural impulse arrives at the presynaptic element, the neurotransmitter-filled vesicles fuse with the presynaptic membrane and release their stores of neurotransmitter into the synaptic cleft (the small space between the pre- and postsynaptic elements). The released neurotransmitter molecules diffuse through the cleft and bind to the neurotransmitter receptors that sit on the membranes of the postsynaptic cells. In response to this binding, channels associated with these receptors open up in the postsynaptic membrane. If the neurotransmitter opens sodium- or calcium-selective channels, the result is usually excitatory, whereas if the neurotransmitter leads to the opening of potassium or chloride channels in the postsynaptic cell, the result is usually inhibitory. For most of the brain's excitatory synapses, the neurotransmitter is glutamate, a derivative of the amino acid known as glutamine. The most abundant inhibitory neurotransmitter in the brain is gamma amino butyric acid (aka GABA), which is also a metabolic derivative of glutamine. There are, however, many different neurotransmitters. Dopamine is a key neurotransmitter involved in brain reward systems and associated with Parkinson's disease. Serotonin is a neurotransmitter involved in appetite and sleep; it is also associated with emotional disorders, such as depression, which is sometimes effectively treated by medications that target serotonin levels. Acetylcholine is the neurotransmitter that all vertebrate motor neurons use to excite muscle cells, and the degenerative muscle disease known as Myasthenia gravis is an autoimmune disease in which antibodies are made against the acetylcholine receptor. And so on.

The formation of synapses is a multistep process that involves many components. On the presynaptic side, the molecular machinery for vesicle production, filling, release on demand, recycling, and refilling must be assembled. On the postsynaptic side, receptors need to be packed tightly together onto a scaffold that holds them in place. Between the presynaptic and the postsynaptic elements, the synaptic cleft has to be enclosed, so that the neurotransmitter that is dumped into the cleft does not diffuse away too quickly. When a growing axon terminal meets an appropriate synaptic partner, things get under way rapidly, as has been studied in tissue culture. Synapse formation begins as soon as the axon of the motor neuron contacts the muscle cell and the two stick together via their corresponding CAMs. In just minutes after making contact, the attachment becomes so strong that when the muscle cell is lifted from the sticky surface of the Petri dish, the growing axon of the motor neuron detaches from the dish and comes with it.

There is little function at the time of first contact, because the synapses are not yet properly working. Signal transmission become vastly stronger and more reliable as the synapse matures. The next proteins that are recruited to the sites of contact are synapse-specific CAMs; the extracellular parts of these molecules, from both the presynaptic and postsynaptic sides, bind tightly to each other, bringing the membranes of the two separate cells even closer together at the site of the future synapse. The intracellular parts of synapse-specific CAMs interact with proteins that begin to assemble the molecular machinery that makes synapses work. As the synapse is being built, many factors, acting like communication signals exchanged between the two main participants, ensure that all the essential components of the mature synapse are built properly. Building a beautiful, reciprocal, synaptic relationship involves a two-way conversation between pre- and postsynaptic partners. For example, the presynaptic endings release molecules that stimulate the production of neurotransmitter receptors on the postsynaptic membrane directly opposite the presynaptic site of release. One of the first such molecules was called "Agrin" because of its ability to aggregate acetylcholine receptors on muscle cells directly opposite the pre-synaptic ending. Many other molecules that are involved in building synapses have since been discovered in the nervous system. Some go in the opposite (post-to-pre) direction, stimu-lating the development of presynaptic machinery directly oppo-site the postsynaptic site.[10]

In 1974, James Vaughn and colleagues at the City of Hope National Medical Center in California began to study synapse formation in the spinal cords of mouse embryos. They looked at thousands of high-magnification images in the electron micro-scope to detect synapses at various stages of their development. What Vaughn and colleagues saw in the images was that new synapses were mostly formed on tips of dendrites. In fact, at early stages of embryo development, most synapses were found on these growing tips, but as time progressed, most synapses became localized to the shafts of the dendrites. From these static images, Vaughn was able to envisage a lively dynamic temporal sequence.[11] First, new synapses are made on the tips of dendritic branches. As these synapses mature, they become stabilized, which helps propel the dendrite on. As the den-drite continues to grow and make new synapses at its tip, it leaves a trail of maturing synapses behind it (figure 6.2).

The successful completion of the early steps of building syn-apses is therefore critical for the characteristic growth of den-drites. If synapse formation is disrupted by mutations in synapse-building proteins, synapses do not get made properly, and so the dendrites of neurons do not continue to grow properly. Time-lapse imaging shows that during their development, the axon terminals and dendrites are extraordinarily dynamic. They are constantly putting out and retracting little branches. Those branches that begin to make new synapses become preferentially stabilized, whereas those that fail to find presynaptic partners are

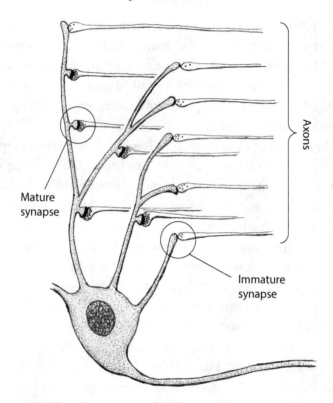

Axons

Mature
synapse

Immature
synapse

FIGURE 6.2. The maturing dendritic tree of a young neuron. New synapses tend to form on tips of growing dendrites, stabilizing these dendrites and allowing them to grow on or branch from there. Synapses mature along the shaft of the extending dendrites.

often retracted, usually within minutes. This hints at a possible competition between the little branches that find a synaptic partner and survive and those that fail to find a partner and are retracted. In the Luo laboratory at Stanford University, recent experiments on the cerebellum of the mouse embryo showed that there is indeed an element of competition in this process.[12] The experimental mice had two kinds of the Purkinje cells in their cerebella. One type had functional synapse-making proteins, and their others did not. The result was dramatic. Not only were the dendrites of Purkinje cells that could not make synapses short and stubby, the dendrites from the cells that could make synapses grew much bigger than usual, as though taking over synapses that would have been made with their neighbors.

Glia Get Involved

The initial ideas and research on making synapses involved just two cells, the presynaptic and the postsynaptic, but these ideas have now been enriched by discoveries showing how glial cells also play critical roles in synapse formation. Ben Barres and colleagues at Stanford University showed that if neurons were cultured in the absence of glial cells, many fewer synapses were made, and those that were made, remained immature and did not work well.[13] Glial cells adhere to forming synapses and secrete factors that stimulate their development. Barres and colleagues used ingenious methods to identify several of these factors. In the absence of one such glial derived factor, called "thrombospondin" (better known for its role in the vascular system), synapses seem to form normally and look anatomically normal in the electron microscope, but they remain inactive, because thrombospondin is necessary for the insertion of neurotransmitter receptors into the postsynaptic cell membrane.

Barres also showed that glial cells are active participants in the pathological processes underlying Alzheimer's and other neurodegenerative diseases. During his remarkable career, Barres made many discoveries that elevated the status of glial cells as active components of brain development and degeneration. Not only was he a key figure in demonstrating the roles of glial cells in many aspects of neural development, but he was also important in another way. He was the first transgender scientist to be elected to the U.S. National Academy of Sciences in 2013. His memoirs, *The Autobiography of a Transgender Scientist*, details his extraordinary life story.[14] Born as Barbara Barres (under which name he published during his early career), he was interested in mathematics and science from a young age. As a student at the Massachusetts Institute of Technology in the 1970s, Barres solved a difficult mathematics problem that had baffled the rest of the class, but the glory was short-lived, as the professor accused Barres of cheating, saying a "boyfriend" must have solved it. Barres transitioned to male in 1997 while a faculty member at Stanford University, and it became apparent to him only then how much he had been discriminated against throughout his previous career. He was surprised to find that for the first time in his life, he could complete a whole sentence "without being interrupted by a man." In 2008, he became chair of the Department of Neurobiology at Stanford University and served in that role until his premature death in 2017.

It is worth taking a moment here to reflect that until synapses are formed in the developing brain, neural communication in the sense of electrical excitability and synaptic transmission has had little or nothing to do with the formation of the brain. Neurons have been born in their vast numbers; they have differentiated into thousands of distinct cell types; and their axons have navigated through the brain, found their targets, sent out branches at topographically appropriate places, found their postsynaptic partners, and started to make synapses with them. This conclusion is also obvious from experiments on laboratory animals that show that the brain develops surprisingly normally when neural activity is silenced during development.[15] Even the brains of human babies that have mutations affecting neural activity look rather normal, though when they are born, such babies may be at great risk of epilepsy or death. It may seem puzzling that so much of the brain could be built by developmental mechanisms that are entirely distinct from the major communication system that the functioning brain uses to do its job. But when one thinks about it, what is the alternative? The establishment of synapses marks a huge turning point in brain development. Neurons can now begin to communicate with one another, which begins a whole new period of brain development. Until this point, building the brain has been a task of construction. But once synapses are established, as we shall soon see, the balance shifts toward demolition.

7

Making the Cut

In which many young neurons compete to make effective synapses, and those that fail commit suicide.

The Death of Neurons

The number of neurons in a human brain is already declining when we are born because many neurons are dying, while few are being generated. In the cerebral cortex, the loss is greatest during the first few years of life; but in many other areas of the brain, the loss has already happened by birth. Roughly half of all the neurons originally generated survive childhood, although the ratio of survival to death varies greatly among distinct neuron types. Why would one ever build a brain this way? Why not make the right number of each type of neuron in the first place? Would an engineer build a computer by carefully putting in far too many microprocessors, connecting them up, and then ripping half of them out? How brains are made seems to me more like the process Michelangelo would use to reveal David hidden inside a block of marble, sculpting away all the unnecessary stone. Or perhaps building a brain is like writing a book, generating a profusion of words, most of which never make the finished manuscript. Or maybe it is more like how a hockey coach might build a team from the players who try out for different positions. But there is no engineer, no sculptor, no writer, no coach involved in selecting which neurons make it through the challenges of brain development. Neurons select themselves for the brain team through do-or-die competitions with one another.

Cell death is not just a feature of the nervous system—it shapes the development of all our organs. It is a standard operating procedure for the building of biological structures. Cell death shapes our fingers by removing the webbing that initially connects them; it removes the cells that connect the lower and upper eyelids, allowing us to open our eyes; it sculpts our immune system, our bones, our guts, our hearts, and our brains.[1]

Cell death occurs in the nervous systems of all animals that have been studied. Even the tiny soil nematode of the species *C. elegans*, which is composed of only 959 cells, produces 1,090 cells during its development, as Sydney Brenner, John Sulston, and Robert Horvitz noticed during their studies of the cell lineages in these animals.[2] They saw that certain cells were destined to die (131 of them, to be precise), many of them in the nervous system. The majority of these 131 cells die soon after they are born. They are programmed to die young! The death of neurons is also a part of the metamorphosis of the nervous systems of insects like moths and butterflies, where many neurons that do important jobs for the caterpillar stages no longer have a function in the flying stages. Frogs also go through a dramatic metamorphosis. As tadpoles, they have a tail that they use for swimming, and they have a type of sensory neuron, called a "Rohon-Beard neuron," in the spinal cord that is involved in the initiation of swimming. Rohon-Beard neurons die

as the tail is resorbed and the animal takes to using its legs. Metamorphosis in insects and frogs is hormonally driven, and it is the
high levels of metamorphic hormones at these critical times that
lead to the death of these neurons. Although humans do not
have a period of metamorphosis as do insects and frogs, we
change just as dramatically in body and brain as we develop.

One might reasonably argue that when building a house, it
would be wise to start with more material than is actually
needed, just in case any of that material is broken or weak. One
may also have to build temporary structures, such as scaffolding, that is subsequently torn down. This is also the case for the
nervous system. There are early transient neurons that reside
above (e.g., the neurons that release Reelin—see chapter 3), and
others that sit just below the cerebral cortex as it is being assembled, and these early neurons provide transient connections between the cells in the different layers of the cortex until
the neurons of these layers are mature enough to establish their
own connections.[3] Like scaffolding when it is no longer needed,
these cells are removed once the cortex has been built. However, the death of cells in the brain as a result of such programmed events accounts for only a fraction of the widespread
death of neurons. Most neuron deaths, as we shall see, are the
result of fierce competitions.

Cell Death and Systems Matching

The first insight into how the brain and its different regions
come to have the appropriate number of neurons came from
Viktor Hamburger and Rita Levi-Montalcini. Hamburger was a
student of Hans Spemann, famous for the discovery of neural
induction (see chapter 1). What Hamburger wanted to find out
was why bigger muscles have more motor neurons to control

them than do small muscles. He was just beginning to do some experiments with chick embryos to investigate the mechanisms behind this proportionate matching, when he lost his job at the University of Frieberg because he was Jewish. He emigrated to the United States, where he was offered a position at the University of Chicago. Hamburger began doing the experiments he had started to do in Germany. He performed microsurgery on the embryos of chickens. He opened the eggs, removed one of the tiny limb buds of the embryos, and then resealed the eggs. Hamburger looked at the spinal cords of the one-legged or one-winged chickens that hatched from such eggs. He found that motor neurons on the missing leg or wing side of the spinal cord were dramatically reduced in number. Hamburger suggested that removal of the limb bud might have removed an inductive signal that stimulated the proliferation of motor neurons.[4]

At about the same time that Hamburger was exploring this phenomenon in Chicago, in Turin, Italy, another scientist by the name of Rita Levi-Montalcini was also interested in experimental approaches to understanding neural development. Also, like Hamburger, she was forced out of her position at her university for being Jewish. So passionate was Levi-Montalcini about her work that she continued her experiments in her bedroom in the family's home in Turin. When Levi-Montalcini read Hamburger's papers, she became fascinated. She not only confirmed Hamburger's results on limb extirpation, but her detailed observations of the spinal cord over the relevant period of development led her to conclude that the reduction in motor neuron numbers on the operated side was due to cell death.[5] When the war ended, Hamburger read Levi-Montalcini's papers and invited her to do a research fellowship in his new lab at Washington University in St. Louis so that they could explore the notion of neuronal death and systems matching more deeply together. Cell death is not easy to see, because it happens so quickly. We now know that as soon as they die, neurons begin to be gobbled up by other cells and are usually gone without a trace within minutes. So Levi-Montalcini counted the total number of living motor neurons at every stage throughout normal chick development. She and Hamburger were astonished and excited to see that the motor neurons in the chick spinal cord are all born by about day 5 of egg incubation, but then their number drops over the next five days.[6]

The natural death of motor neurons even in normal chick embryos combined with the fact that survival seems to depend on the target suggests how the developing nervous system selects its team of motor neurons. Only enough muscle cells exist in any muscle to accommodate the survival of about half of the incoming motor neurons, so the motor neurons compete for survival. Hamburger and colleagues tested this hypothesis by surgically transplanting an extra limb bud onto a chicken embryo. Such chickens hatched with three legs, two on one side and one on the other. The side with just one leg had the normal number of motor neurons, while the side with the extra leg had many more, not because any extra motor neurons had been generated but because fewer had died.[7]

These experiments indicated that to match the population size of a pool of motor neurons to the muscle that these neurons innervate, a superabundance of motor neurons is first generated, and then this excess is trimmed away to leave an appropriate number. If there are more muscle cells to make synapses with, more motor neurons will survive. Such effects can percolate through the developing nervous system to regulate the numbers of the different types of neurons. For example, the size of the pool of surviving motor neurons may have effects on the sizes of the populations of neurons that are presynaptic to them. This early work by Hamburger and Levi-Montalcini on systems matching by cell death has been reinforced by many more recent studies that have found similar effects in other regions of the nervous system. Target-dependent neuronal survival is a rather ubiquitous feature of brain development.

Neurotrophic Factors

Hamburger and Levi-Montalcini wondered what it was about muscle cells that could keep motor neurons alive. They considered the possibility of a survival factor that the target cells feed to their innervating neurons. They considered, for instance, that there might be a limited supply of this survival factor, enough to nourish only half of the innervating neurons. But did such a survival factor really exist? One way that cell biologists search for factors is to test various cell lines to see whether any of them secrete the factor they are looking for. So, various cell lines were injected into the limb buds of chick embryos. One of these, a human sarcoma cell line, had a truly profound effect on sensory neurons. It stopped their cell death, so that all the sensory neurons that were generated also survived, and it also stimulated the tremendous growth of these neurons. It had what is known in cell biology as a "trophic" (Greek for "nourishing") effect. Hamburger and Levi-Montalcini named the active component "nerve growth factor" (NGF).[8] The final purification of NGF, the first discovered neurotrophic factor, was done as collaboration between Levi-Montalcini and biochemist Stanley Cohen, who was also at Washington University.[9] They were jointly awarded the 1986 Nobel Prize for this work. Many think Viktor Hamburger deserved a share of that prize.[10]

A great deal of experimental work has been done on NGF since its initial discovery. It is a peptide (i.e., a short stretch of amino acids), and it is essential for the survival of sensory neurons and the neurons of the sympathetic nervous system. If these neurons are put into a Petri dish without NGF, they quickly die. Receptors for NGF are found on the axon terminals of NGF-dependent neurons. When NGF is released from target cells, it binds to these receptors. The receptors bound to NGF then become internalized and are transported back along the axon to the cell body, where they give their vital message to the nucleus: "Survive and grow!" If enough such messages are delivered, the cell survives and grows. If there are insufficient activated receptors, the neuron retracts its axon and dies.

One of the biggest surprises of the early research on NGF was that it seemed to have no effect whatsoever on the survival of motor neurons, even though their numbers are affected in the same way as sensory neurons by limb-bud removal or addition. This inspired the search for other neurotrophic factors. Sure enough, other trophic factors were found. Although each neurotrophic factor clearly affects certain populations of neurons, many types of neuron rely on more than one neurotrophic factor. For example, retinal ganglion cells rely on three different neurotrophic factors to regulate their survival and distinct aspects of their axonal or dendritic growth. Motor neurons depend on a combination of at least two trophic factors that are distinct from NGF, at least one of which has still not been identified.

The potential of neurotrophic factors to keep neurons alive and stimulate their growth makes them exceptional candidates for the treatment of a multitude of neurodegenerative disorders as well as brain and spinal cord injuries. Medical research on neurotrophic factors is very active, and much progress has been made in our understanding of how they work. However, translating this research into effective therapy for humans has been challenging, in a large part due to the blood-brain barrier, which prevents proteins and peptides, like NGF, injected into the blood stream from getting to the brain cells that need them.

Apoptosis

The neurons that are starved of neurotrophic factor do not die gently. They commit cellular suicide by digesting their own proteins and chewing up their own DNA, a phenomenon known in cell biology as "apoptosis" (from the Greek for "dropping off"). Apoptosis is an active process that requires the cell to first make and then engage the machinery that will demolish the cell.

Robert Horvitz and colleagues searched for and found cell-death mutants in the nematode *C. elegans*, and they found some mutants in which all 131 cells that were programmed to die soon after their generation survived instead.[11] Many of these 131 cells are the sisters of cells that become functional neurons, and if they are kept alive, as they are in cell-death mutants, they become "undead" neurons. Some of these undead neurons share the same neuronal fate as their sisters, and others develop abnormally and make strange synaptic connections, with negative consequences for neural function. If neuronal cell death is prevented in fruit flies, the flies begin to wriggle as little larvae at about the normal time, but most do not survive to hatching stages.

The genes that are faulty in these mutants reveal the trigger mechanism for apoptosis. Indeed, the proteins made by these "cell-death" genes activate a cell-death pathway that can be analogized to the action of firing a gun. Pulling the trigger releases a spring-loaded hammer that hits the back of a shell, causing the gunpowder to explode. The cell-death proteins are arranged in a cascade such that the first activates the second, which activates the third, and so forth. It is the final step, activating the digestive enzymes, that is the explosive one that kills the cell, and from which there is no return. The question then becomes: What exactly causes a cell to pull the trigger that leads to its inevitable death?

Martin Raff, a scientist at the MRC Laboratory of Molecular Biology in London, realized the connection between the availability of neurotrophic factors and apoptosis.[12] In the vertebrate nervous system, apoptosis is the result of not obtaining enough neurotrophic factors. Raff proposed that every developing neuron is on the verge of apoptosis, holding a gun to its head, so to speak, with its finger ready to pull the trigger if it does not get enough neurotrophic factors. Whether activated by a lineage-based program or neurotrophic factor starvation, the cell-death pathway is basically the same in nematodes as it is in humans. It uses the same molecular components—the cell-death pathway gene products—for the trigger mechanisms.

Apoptosis is at the heart of many neurodegenerative diseases, where the cell-death pathway is activated. And it is also at the heart of many cancers, such as lymphomas, where the cell-death pathway fails to be activated, so that aberrant cells do not die when they should. Various circumstances can initiate apoptosis, like inadequate neurotrophic factors, and cells use various safety mechanisms to protect themselves from accidentally pulling the trigger. The clinical relevance of understanding and being able to control apoptosis is therefore huge. But because apoptosis is evolutionarily ancient and works more or less in the same way in all animal cells, a drug that blocks the apoptosis pathway might save some cells but allow other cells to become overabundant, while activating the pathway might help kill cancer cells but also kill other cells that are essential for survival. Research and clinical trials continue on different pharmacological approaches to control the cell-death pathways.

Activity and Death

The battle for neurotrophic factors is only one of the battles that a neuron must win to survive. As well as establishing output synapses with their target cells, neurons also need to have successfully made synapses with the cells that innervate them. Some of the experiments that Rita Levi-Montalcini did in her bedroom laboratory in the 1940s, before she went to work with Viktor Hamburger, had shown that the ear was necessary for the survival of neurons that receive direct connections from the auditory nerve.[13] Subsequent experiments on different brain regions have shown that receiving synaptic input is also critically important for the survival of many types of neurons. But in this case, what the presynaptic cells provide is usually not a neurotrophic factor but simple electrical activity. Just as artificial electrical stimulation of muscles can counteract muscular atrophy, artificial electrical stimulation of neurons through implanted electrodes appears to have salvaging or protective effects on neurons that have been experimentally deprived of input. Direct electrical stimulation has also been hugely successful in treating Parkinson's disease patients, and positive effects of such stimulation have also been seen with Alzheimer's and other dementias.[14]

The importance of using activity for regulating neural survival is highlighted in the exquisite balance between excitation and inhibition in the brain. Almost all neurons in the brain have some excitatory and some inhibitory inputs, and the balance must be adjusted so that the brain is neither too active nor too quiet. The proper ratio of excitatory and inhibitory neurons is critical for this balance. A pathological imbalance between excitation and inhibition during brain development in humans is thought to be the underlying cause of several forms of epilepsy and autistic spectrum disorders.

During normal development, many inhibitory neurons migrate into the cerebral cortex. The survival of these neurons depends on their receiving effective excitatory input from other cortical neurons. This, however, creates a feedback loop because these inhibitory neurons inhibit the neurons that excite them, the neurons whose activity they depend on for survival. The reduction in excitatory activity thus limits the survival of the inhibitory neurons, for as the overall amount of cortical activity falls, it eventually reaches the threshold at which there is just sufficient activity to maintain the remaining inhibitory neurons whose death at this point would lead to a rebound of excitation. This thermostat-like feedback mechanism,[15] which initially sets the general level of neuronal activity in the cortex, again works on the basis of self-corrective cellular suicide. Those inhibitory neurons that do not get enough excitation commit apoptosis.

Following the period of great cell death, the neurons that survive—those that have made the cut and have been selected to join the brain team—appear to become less reliant on synaptic connections. If we return to the early work of Rita Levi-Montalcini on removal of the developing ear, she showed that the death of auditory neurons in the chick embryo brain happens at about day 10 in development (about halfway through its time in the egg), when the neurons are becoming innervated from the auditory nerve. If the ear is removed a few days later, however, the neurons do not die. There is thus a crucial period. It is during this period that neurons find that they have received and maintained enough effective connections and should survive. Perhaps once the nervous system has rid itself of all the unnecessary neurons, it strives to keep alive all the remaining ones—as they are never replaced.

Birth is the start of the period of brain development that interests most people, but it comes near the end of this story, which followed the neurons of the brain from their earliest origins as a set of a thousand or so neural stem cells in the gastrula. We saw them proliferate and become the myriad different types of information-processing units that inhabit the brain. In this chapter, we witnessed the phase of brain development in which many neurons are eliminated. Some neurons have done their jobs and are no longer needed, but most neuronal deaths seem to occur because neurons must compete for the right to survive and become lifelong members of the brain team. The neurons that do not make the cut eliminate themselves through apoptosis, the suicide-like process of self-digestion. Most neurons that do make the cut are here for a lifetime, but just because a neuron survives the great cull does not mean that its development is complete, for at the synaptic level, the brain is overconnected. A great period of refinement is about to commence.

8

The Period of
Refinement

In which synchronized patterns of electrical
activity are used to refine the synaptic networks
in the brain during critical periods of prenatal
and postnatal life.

Pruning the Brain

Though the number of neurons in the brain is decreasing at
birth, the number of synapses is rising. At a microscopic level,
one envisages growing tangles of dendritic branches and axonal arbors intertwining thickly with one another and filling
any spaces with new synapses. Synaptic numbers continue to
rise in the cerebral cortex of humans until about four years of
age. But then, the number of synapses begins to drop. As axons
and dendrites stop growing, fewer new synapses are made, and
many of the existing synapses between neurons are eliminated.
This greatly refines the wiring diagram of the brain. This is the
period of neural development during which the brain makes
major adjustments to its circuitry and begins to fine-tune itself.

As synapses are made, the brain to begins to fire up, and as it does so, it reveals that the process of wiring has been largely successful. One might be tempted to say, "too successful." Most target cells have more synaptic inputs than they seem to need. For example, consider the synapses made between motor neurons and muscle cells. In our postnatal bodies, each muscle fiber is innervated by a single motor neuron. But initially, each muscle fiber is poly-innervated (i.e., it make synapses with up to five different motor neurons). How is refinement achieved in this case? One of the world's most potent snake toxins, known as alpha-bungarotoxin, which is harvested from the many-banded krait of Taiwan, has helped to answer this question. If a many-banded krait bites you, it injects the toxin, which finds its way into the synaptic clefts in the synapses between your motor neurons and your muscle cells. It binds tightly to the acetylcholine receptors there, making them blind to the acetylcholine released by your motor neurons. Muscles (including your diaphragm muscles) no longer respond to neural stimulation—you try to move, to breathe, but you are unable.

A key question for developmental neuroscience was whether blocking synaptic transmission had any effect on the formation of synapses. Alpha-bungarotoxin provided a way to ask this question at the synapses between motor neurons and muscle cells. In early experiments, the toxin was given to chick embryos during the period of synapse refinement. The synapses between the motor neurons and muscle cells of the immobilized embryos were then compared with those of control embryos. The anatomy of the synapses that formed looked normal, even at an electron microscopic level. However, physiological and anatomical studies of muscles that have been grown in the presence of alpha-bungarotoxin showed that muscle fibers remained poly-innervated. Such experiments indicate that synaptic function is essential for synapse elimination.[1]

How does synaptic activity eliminate all but one motor neuron from each muscle fiber? The rule seems to be: "Keep the most effective one." Indeed, experiments on individual muscle cells that were innervated by two motor neurons show that the activation of one axon strengthens its synapses at the expense of the synapses from the nonactivated axon. The more active and effective synapse is rewarded every time it evokes a response in the muscle, so it grows and takes over territory, while the muscle appears to punish any synapses that did not contribute to its activation. In this way, the weaker synapses are actively removed, leaving just one motor neuron axon terminal on each muscle cell (figure 8.1). The molecular natures of these reward and punishment signals are still unknown.

Like muscle cells, many neurons in the brain receive too many synapses from too many presynaptic cells at first. The brain at this stage is said to be in a state of "exuberant connectivity." Brain neurons, like muscle cells, go through a process of synapse elimination driven by activity patterns that serve to eliminate those synapses that are not working as effectively as others. For the presynaptic cell or postsynaptic cell, synapse elimination may often lead to the loss of an entire branch of an axon or a dendrite. In an analogy to gardening, this is called "pruning." Branches that have made the most effective synapses are saved, while the weaker branches that lose their battles are pruned away by glial cells, which gobble them up. In the end, each axon often ends up with more synaptic territory than it had at the beginning, but it is focused on fewer postsynaptic dendrites. The molecular mechanisms underlying synapse elimination are under active investigation and may have implications for Alzheimer's and other neurodegenerative diseases in which synapse loss is a key problem.

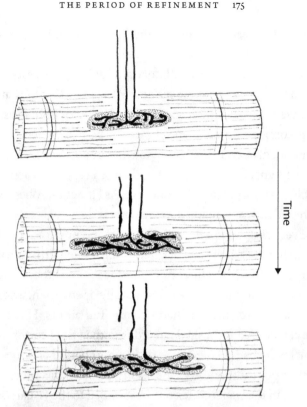

FIGURE 8.1. Synapse elimination over a course of three time points. Three motor neurons originally make synapses with a muscle fiber (top panel). The muscle fiber is the large cylindrical structure, and the axons of the three motor neurons run down into the muscle, thickening where they are making synapses in the stippled area of the muscle. In the middle panel, the synaptic contact made by one of these axons (leftmost) is eliminated, and the synaptic contact made in the bottom panel by a second motor neuron (center) is eliminated, leaving the muscle fiber innervated only by the rightmost axon (bottom panel).

Critical Periods

David Hubel and Torsten Wiesel were awarded the Nobel Prize in 1981 for the amazing discoveries they made about the visual cortex.[2] Working first at Johns Hopkins University and then at

Harvard, they explored the way that single neurons of the visual cortex respond to stimuli. One of their first exciting findings was that in mammals like us, with forward-looking eyes, most neurons in the visual cortex are binocular, meaning that they are responsive to visual stimuli coming from either or both eyes. The axons coming into the cortex, however, are not binocular. They are monocular: each incoming axon fires impulses in response to a visual signal that comes from only one eye, and its activity is unaffected by input into the other eye. As Hubel and Wiesel were interested in how visual experience affects the visual cortex, they did a very simple experiment with profound implications. They sewed shut the right eyelid of kittens shortly after birth, and then, when the cats reached three months of age or later, they removed the stitches, so that both eyes were fully open. Then, a year or more later, after the cats had become full adults, Hubel and Wiesel recorded from single neurons in the visual cortex of the animals who had been exposed to this short period of early visual deprivation to just one eye. The results were striking. Most of the neurons in the visual cortex only responded to the visual stimuli presented to the left eye. Indeed, almost all the neurons in the deprived cat brains were dominated by the left eye. Hubel and Wiesel put a patch over the left eye of an adult cat to see whether it could navigate with its deprived right eye. The cat seemed blind with just the right eye; it walked into walls and fell off platforms.

In stunning contrast to the effects of early deprivation, deprivation during adulthood did not have dramatic effects on the visual cortex. Hubel and Wiesel sewed shut the right eyelids of adult cats for a year or even longer and found no significant changes in the cortex. As soon as the eye was opened again, the visual cortex got the signal, the neurons in the visual cortex were just as responsive and binocular as before, and sight through the deprived eye was normal. For the deprivation to have a significant effect on the cat's visual cortex, it had to occur within the first three months of life—this period is critical for the development of the cat visual cortex. Hubel and Wiesel also found that when kittens are first born, most of the neurons in the visual cortex are already binocular and respond to visual stimuli coming from either eye. This suggested to them that during this critical period, connections from the deprived eye are lost in the cortex while those serving the open eye remain and expand. They concluded that the physiological defect in the deprived kittens represented a disruption of connections that were present at birth.

Do we humans have such a critical period for visual cortex development? We do. In fact, Hubel's and Wiesel's work on visual deprivation was inspired by clinical observations that children with congenital cataracts can have permanent visual deficits even if clear sight is restored during adulthood. Similarly, if a child cannot focus well through one eye, they often develop what is known as a "lazy eye," an eye that does not contribute much to vision. Lazy eye can sometimes be cured by simply wearing corrective lenses. Early treatment works best, but if such treatment does not start by the time the child is about eight years old, it will be too late—the visual cortex will become permanently dominated by the nonlazy eye.

Are there critical or sensitive periods for other senses, or other functions of the brain? Again, the answers are yes and yes. Critical periods seem to exist for all sorts of cognitive functions in animals and humans. A wonderful example of a sensitive period is that for accurate sound localization in barn owls.[3] We know that barn owls are able to locate prey, such as voles, even in total darkness, by tracking their sounds. In experiments that are similar to visual deprivation, Eric Knudsen at Stanford University plugged one ear of young barn owls. This caused the birds to make errors in locating the sources of sounds in the dark. However, over the course of just days, the representation of auditory space in the brain readjusted, and the animals regained their ability to strike accurately in the dark. If the plug was then removed before the animal was two months old, the owl could quickly adjust again, but if the plug was removed after the animal was two months old, recovery became greatly protracted.

The most famous example of a critical period is from one of the founders of the field of ethology, Konrad Lorenz (1903–1989). Lorenz made the dramatic observation that during a brief period of only a few hours after hatching, greylag geese form stable attachments to the first object that they see moving.[4] This is known as imprinting. Once the imprinting period ends, the preference for the imprinted stimulus remains strong—young geese that had imprinted on Lorenz during their first day followed him rather than their own mother. Zebra finch nestlings become imprinted on their mothers; if raised by a foster mom of another species (say, a Bengalese finch), they will court that species when they grow up, no matter how many sexy zebra finches they see.[5] Baby zebra finches, like other songbirds, also imprint on the songs of their fathers. Learning and memorizing the song happens first, during a limited period early in development. If male birds do not hear their father's song, or one very like it, early in their lives, they can never learn to sing it properly themselves. A second critical period opens when the bird is old enough to practice the song aloud, and if the bird is deprived of hearing itself sing during this period, again proper song is never acquired. This sounds a little like how humans learn to speak a language (more in chapter 9).

Synchrony

One of the most surprising aspects of Hubel's and Wiesel's visual deprivation experiments was the discovery that a competitive process drives it. Having seen the loss of connections from the deprived eye, Hubel and Wiesel had a "use it or lose it" idea in mind. They anticipated that closing both eyes of a kitten throughout the critical period would compromise the connections from both eyes. So, they were a little bewildered when they discovered that if both eyes of a kitten were closed throughout the critical period, most of the cells in the visual cortex continued to respond to both eyes. Amazingly, the kittens' binocularity seemed almost normal! If both eyes were equally disadvantaged, neither won and neither lost; they were each able to hang on to their postsynaptic neurons in the cortex. It seemed that it was the imbalance of activity that produced a loser and a winner. The synapses that have been most active and successful in driving their postsynaptic partners are the winners, and they take over the synaptic territories previously held by the losers.

The next question was: What if both eyes worked well, but never together? To ask this question, Hubel and Wiesel put a patch over one eye for a day and then the other eye for the next day, alternating eyes each day throughout the critical period, so that both eyes had a similar amount of visual stimulus but never saw the same image at the same time. Again, they sampled the activity patterns of hundreds of neurons in the visual cortex of such cats. What they found in this case was that half of the neurons in the visual cortex responded to input through one eye, and the other half responded to input through the other. No neurons were binocular. This is similar in some ways to a condition in humans known as strabismus, where the two eyes are not properly aligned. If a baby has strabismus, and the eyes are not straightened out during early childhood, the result is the permanent loss of binocular neurons and the consequent loss of binocular depth perception. The concept arising from these experiments is simply that for neurons of the visual cortex to remain binocular through the critical period, both eyes must see the same things at the same time.

Cortical cells that receive inputs from synchronously active neurons do not have winner and loser inputs: they retain both sets of connections and remain binocular. These results are similar in many ways to the results described above for the strategy that muscle cells use to refine their innervation. Like closing both eyes, if the inputs from motor neurons to muscle cells is experimentally silenced by alpha-bungarotoxin, the poly-innervated state is maintained. Muscle cells also remain poly-innervated if the activity of the motor neurons that innervate them is experimentally synchronized, which is similar to what happens in the visual cortex when both eyes are open throughout the critical period preserving binocularity. Synchronization is the key.

Insight into the concept of neural synchrony and how it could be used as a basis for modifying the strength of synapses came first from the neuropsychologist Donald Hebb (1904–1985) of McGill University. Hebb was interested in synaptic theories of learning. He formulated what is now known as Hebb's rule, which he stated thus: "When an axon of cell A is near enough to excite a cell B and repeatedly or persistently takes part in firing it, some growth process or metabolic change takes place in one or both cells such that A's efficiency, as one of the cells firing B, is increased."[6] Hebb's rule can be roughly equated to the catchphrase "cells that fire together wire together," though Hebb's rule has more subtlety: only if cell A's firing is important to cell B's firing would this lead to the strengthening of the synapse. The strengthening of one synapse often leads to the weakening of another, so a corollary of Hebb's rule is that synapses that do not take part in the firing of the postsynaptic cell are weakened. The catchphrase for this corollary is "stay in sync or lose the link." With this in mind, we can now go back to the Hubel and Wiesel experiments. Imagine two cells, neuron L and neuron R (for left and right). Both

R and L make synapses with a neuron V in the visual cortex. If cell L causes V to fire better than R does, L's connection is strengthened, and R's connection is weakened. But if L and R see the same things at the same time, their inputs to V will arrive at the same time, and both synapses will stay in balance and be retained. Hebb's rule and its corollary seem to provide an excellent basis for thinking about all the results of the visual deprivation perturbation experiments that define the critical period for the refinement of synaptic connections in the visual cortex as well as the refinement of muscle cell innervation by synapse elimination.

Test Patterns in the Womb

The first synapses in the human nervous system are made in the spinal cord long before birth, at just five weeks of gestation. The spinal circuitry then begins to be refined during the first and second trimesters. The circuits of the hindbrain and midbrain begin to be built and refined next, during the second and third trimesters. The forebrain, particularly the cerebral cortex, is the last to develop and the last to be refined, happening postnatally to a large extent. It is not until we sense the world outside the womb that the development of the cerebral cortex can be effectively tuned to specific features of the external world. A baby's brain may be primed to learn a mother's face, but the detailed information about the particular face of the baby's mother is not available until the baby opens its eyes and sees her. Real-world visual signals do not reach the brain of a fetus, protected deep within the warm darkness of the womb, with its eyes closed. Yet even at this early stage, the brain is being sculpted by the same sorts of Hebbian mechanisms that continue to be used to learn about the outside world postnatally.

In 1991, Marcus Meister, Rachel Wong, Dennis Baylor, and Carla Shatz at Stanford University discovered waves of synchronized activity in the retinas of fetal cats and neonatal ferrets.[7] These waves resembled those that were first broadcast during the 1986 World Cup competition in Mexico and became known as "Mexican" waves. In a Mexican wave, successive sections of the crowd stand up, raise their arms, lower their arms, and then sit down again. Similarly, in the retina, neighboring retinal neurons become transiently active at the same time as their neighbors, creating waves of electrical activity that travel in different directions across the retina. These patterns of activity are somewhat similar to the way that the retina would respond to slow-moving blurry images of the real world. Yet all this is happening in the complete darkness of the womb, the eyes are closed, and the photoreceptors that sense light have not even begun to work. These spontaneous waves of retinal activity are happening long before vision is even possible.

Retinal ganglion cells are the output neurons of the retina. They send their axons along the optic nerve and into the brain. Retinal ganglion cells are participants in these prenatal Mexican waves just at the time when their axons are making synapses with their target neurons in the optic tectum. Up to this point, the retinal ganglion cells' axon terminals started to make synaptic connections in the optic tectum based on chemical gradients of Ephrins (see chapter 6). The topographic mapping is good but not yet fully refined. The retinal waves mean that the closer one retinal ganglion cell is to another on the retinal surface, the more likely it is to be active at the same time. Hebbian mechanisms use this information to make the map functionally precise. It is a beautiful solution to the problem of how to sharpen the visual map before any visual input is available. It is the developing brain's way of running its own test patterns to fine-tune the connections and sharpen the image.

Carla Shatz and her colleagues at Stanford showed that these pre-visual retinal waves are also involved in sorting out the connections between the left and right eyes in the brain. In all mammals, retinal ganglion cells from both eyes send axons to a thalamic brain region called the "lateral geniculate nucleus" (LGN). The LGN is a layered structure built a bit like a stack of pancakes, each with its own map of the visual world. In an adult, each pancake in the stack receives inputs from only one eye or the other, but in early stages of brain development, when young retinal ganglion cell axons first arrive in the LGN, they make synapses in both right and left layers. There is no correspondence in the patterns of the retinal waves in the left and right eyes as the retinal waves of activity arise spontaneously, rather than being driven by real images. The Hebbian corollary works to ensure that inputs that are not in synch lose their links. The result is that the stacks sort themselves neatly into left eye-dominated and right eye-dominated layers.[8]

Sorting the LGN into left and right layers that are monocular happens well before birth in humans, long before any real image of the world would have come into a baby's eyes. By the time a newborn baby first opens its eyes, the neurons of the LGN are firmly monocular, and their critical period is over. These monocular LGN axons send their axons into the visual cortex. The sorting out of these LGN axon terminals within the visual cortex begins with the opening of the eyes and seeing the world for the first time. Real vision then allows the same Hebbian mechanisms of synchronicity to sharpen various features of the real image, such as binocularity. As Hubel and Wiesel showed (see above), both eyes must be open during the critical period to preserve binocularity in the visual cortex.

It is, of course, not only the visual system that is tuning up itself before birth. Patterns of spontaneous activity have been observed throughout the developing nervous system. There is good evidence that spontaneous patterns of neural activity in the womb are involved in tuning the auditory system, the motor system, the cerebellum, the olfactory system, and many other developing neural circuits in the brain.[9]

Tuning In and Tuning Out

We have a gut understanding of childhood as a particularly impressionable age, which the 2020 Nobel Prize winner in Literature, Louise Glück, put so succinctly in her poem "Nostos":[10]

> Fields. Smell of the tall grass, new cut.
> As one expects of a lyric poet.
> We look at the world once, in childhood.
> The rest is memory.

Spontaneous patterns of activity, like the internally generated retinal waves, begin to sculpt the circuits of the brain long before birth. However, critical periods of circuit formation extend into postnatal life, especially in the cerebral cortex, so that the world outside the womb participates in fine-tuning the brain.[11] Plasticity in the cerebral cortex diminishes with age. As we leave childhood, we seem to shut down certain critical periods and put the "brakes" on plasticity. We do not know why plasticity slows down. There are various possible explanations: for example, if tuning in means the adjustment of trillions of synapses so the system works optimally, it could tend to stabilize in such a configuration. As cells tune themselves in so that they operate in an optimal range for the world they are becoming familiar with, the circuitry becomes more locked in, and in a way, more tuned out to the changing world.

Structural studies in the brain appear to hold other clues to the secrets to shutting down plasticity. For example, in several areas of the cortex, one can find a type of inhibitory neuron that seems to accrue extracellular material around itself. In the microscope, this material looks like chain mail wrapped around the neuron, constraining further changes in its basic shape. Meanwhile the glial cells known as oligodendrocytes are wrapping up axons in myelin, and astrocytes are encapsulating synaptic contacts. All these processes, and indeed several others, have been proposed as parts of the basis for closing down critical periods and limiting plasticity.

Neuroscientists have been searching for ways to reopen critical periods. Would it be possible, for example, to reopen a critical period for language acquisition so that as adults, we might be able learn a new language like a child does? Similarly, psychologists may wonder whether the damaging effects of an adverse experience or of social deprivation during childhood could be counteracted by reopening a critical period in the adult brain. If we could identify and understand the neural circuits that have been adversely affected or neglected during critical periods of brain development, perhaps then we could also find ways to restore more normal function to them.

Learning

Be heartened: plasticity of synapses never completely stops. Even a neuron in the adult visual cortex with its thousands of synapses arranged along its dendrites, a neuron that now functions in a superbly efficient way, gathering and processing pieces of visual information, can keep making small adjustments to keep itself tuned as the years roll by because every one of those synapses is still a bit modifiable. Each of the hundreds of trillions of synapses in the brain can become stronger or weaker throughout life. The persistence of this type of synaptic plasticity into adulthood is what allows learning. But the large adjustments that were made so rapidly in youth are sometimes impossible for adults.

Learning is a kind of persistent plasticity in the brain, and it is easy to view this as an extension of the fine-tuning processes discussed above. In fact, Hebb originally formulated his rule to provide an explanation for associative conditioning in adult animals, the type of learning that Ivan Pavlov studied in dogs. A dog salivates to a bell that has been rung immediately before the presentation of food. Food has always been enough to cause the dog's saliva to flow. However, if the dog repeatedly hears a bell just before his food is presented, synapses that associate the sound of the bell with the presentation of a food reward become strengthened, and soon the bell alone will cause the dog to salivate. With Hebbian mechanisms in mind, researchers around the world began to investigate the neural basis of learning by focusing on the ways that synapses could be strengthened or potentiated. Thousands of scientific papers have now been published about cellular learning mechanisms. Below is the briefest overview of these mechanisms, given in the context of brain development.

Memory circuits that associate one thing with another, like food with the sound of a bell, often work in the following general way. Presynaptic neurons carry an array of inputs. If two or more of these inputs are synchronously active and contribute to firing the postsynaptic cells, these inputs become strengthened, or potentiated. In a typical synapse in the vertebrate brain, the excitatory presynaptic component releases the neurotransmitter glutamate when it fires. The postsynaptic cell has two main types of receptors for glutamate. One type opens ion channels that begin to excite the postsynaptic neuron, but the other type, known as NMDA (N-methyl-D-aspartate) receptors, do not open their ion channels unless the postsynaptic cell has first been sufficiently activated by the non-NMDA glutamate receptors. It is only when the postsynaptic cell is effectively stimulated by non-NMDA receptors that the NMDA receptors themselves will become active. Thus, the NMDA receptors act as "coincidence detectors" between presynaptic glutamate release and postsynaptic cell activation. When NMDA receptors are active, they open calcium channels in the membrane of the postsynaptic component of the synapse. The local influx of calcium is a signal to strengthen that synapse by inserting more receptors for glutamate and setting in motion a process of local growth. In addition, feedback to the presynaptic element causes it to become more efficient and larger. At first, the synapses that signal the sound of the bell are not enough to activate the postsynaptic neurons that lead to salivation, but the food is always enough to excite these neurons. If the bell synapses release glutamate when the salivation neurons are already active due to the presentation of food, the bell neurons can then activate NMDA receptors, and this strengthens their synapses. Soon the bell synapses by themselves are strong enough to activate the salivation neurons. Such molecular mechanisms are thought to be at the heart of how most neurons store new information in the modification of synapses. Perhaps it is not a surprise, then, to learn that these same NMDA receptors are also at the heart of the mechanisms involved in the periods of refinement. For example, if NMDA receptors are pharmacologically blocked during the critical period of visual development, binocularity is maintained in the visual cortex even if one eye is deprived of vision throughout this period.

The causality implicit in Hebb's rule means that if cell A fires later than cell B, it could not have caused cell B to fire, and so the synapse should not be strengthened; it should perhaps even be weakened. I was lucky to be a colleague of Mu-Ming Poo, when he was working at the University of California, San Diego. One day Mu-Ming turned up at my lab and asked whether we could show him exactly where the optic tectum was in an embryonic *Xenopus* brain. Once he and the scientists in his lab learned how to record from cells in the *Xenopus* optic tectum, they were able to use a three-neuron system (two presynaptic retinal ganglion cells and one neuron in the optic tectum that was postsynaptic to both) to define a window of synchrony necessary for such synaptic strengthening and weakening. If the presynaptic cells fire within tens of milliseconds before the postsynaptic cell fires, their synapses will be strengthened or potentiated. But if the presynaptic cell fires just slightly after the postsynaptic cell, the synapse is weakened or depressed.[12]

Our brains, like the brains of many animals, change throughout life. We do not change who we are, but our brains, like all our organs, continue to change within us. The next question is then: "Who exactly are we, and how is this encoded in our brains?" Chapter 9 sheds some light on this big question from the perspective of how the brain is made.

9

Being Human and Becoming You

In which humans evolve human brains, and we find
that the mechanisms that make brains human
operate in ways to ensure that every human being
has a unique mind.

Does Size Matter?

The brains of human beings are different from those of every
other species of animal because all species' brains have been
tuned to their lifestyles through millions of years of evolution.
A spider's brain is geared to weaving webs and catching flies, a
fish's brain is tuned for a life in the water, and a human brain is
geared to human affairs. The previous chapters of this book
emphasized many similarities between animal and human
brains that are rooted deep in evolution and in the biological
mechanisms used to build nervous systems. Here, we focus on
differences between human and animal brains, how these differences arise, and what these differences might mean for us as
individuals.

One of the things that neuroanatomists who are interested in evolution like to compare is brain sizes.[1] The average adult human brain weighs about 1.5 kilograms—pretty big, but not the biggest. The brains of African elephants weigh about 5 kilograms, and the brains of sperm whales weigh about 8 kilograms. Mice have tiny brains, less than a gram. The smallest known mammalian brain, weighing in at just 64 milligrams, belongs to the Etruscan shrew. The smallest of all known brains on the planet belongs to a species of parasitic wasp, *Megaphragma mymaripenne*, whose whole body is the size of some protozoans. In these wasps, neurons that make up the brain lose their cell bodies and nuclei soon after they are generated. The wasp's tiny brain is thus made up almost entirely of axons and dendrites, just the living wires and their connections.[2] This is enough to allow them to live busy but very short lives, flying in search of prey and potential mates.

The human brain accounts for about 2% of the average adult body weight. So, perhaps we have the biggest brains in proportion to our bodies? But no, we do not have the highest brain-weight to body-weight ratio. Small mammals tend to have high ratios, while large animals have low ratios. This is largely explained by a scaling relationship between brain and body mass: animals that are ten times larger than other animals tend to have brains that are only six times larger. Why this particular scaling occurs is a matter of speculation among comparative biologists. If, however, one takes this scaling relationship into account and asks whether a human brain is larger than expected for its body size, the answer is, finally, yes. Our brain is about 10 times larger than expected for a mammal of our size. Human brains are 4 times as big—and we have about 4 times as many neurons in our brains—as our closest living relatives, the chimpanzees, whose body size is similar to ours.

Although we do not have specimens of preserved brains from our hominin ancestors, their skulls provide some information. The braincase of the skull can be used as a proxy for the size and shape of the brain, especially the cerebral cortex. Studies of braincases points to an expansion of the cortex in *Australopithecus*, the earliest known hominins, who appeared in Eastern Africa about 4.2 million years ago. Lucy, the most famous australopith, had a brain that was about a third the size of a modern human brain, which is a bit bigger than a modern chimp's brain. Lucy walked on the ground on two feet, leaving her hands free to grasp tools. Being bipedal, she would have been able to raise her eyes over tall grasses and survey the distances for food or danger. A second cortical expansion is seen with the emergence of *Homo erectus* ("upright man") about 2 million years ago, with brains about half the size of a human brain. *Homo erectus* are thought to be the first who used fire, who worked together in large communities, who ventured out to sea, and who made art. The final cortical expansion is seen in our likely ancestors *Homo heidelbergensis*, whose brains were essentially human in size. *Homo sapiens* (modern humans) are thought to have diverged from *Homo heidelbergensis* about 300,000 years ago. Since the origins of *Homo sapiens*, the brain cases of our species have hardly changed.

Neanderthals are also thought to have emerged from *Homo heidelbergensis* and coexisted with humans until about 35,000 years ago, when they went extinct. Their brains were slightly bigger than those of *H. sapiens*, but a difference of this magnitude is to be expected, simply because Neanderthals had bigger bodies than we have. Analyses of shapes of the braincases show that Neanderthal brains were slightly more elongated than those of *H. sapiens*, and this has allowed some speculative suggestions about how human brains and mental processes may be somewhat distinct from those of the Neanderthals. For example, the occipital lobes, which are involved in vision, are relatively larger in Neanderthals than in modern *H. sapiens*, so visual processing may have been superior in Neanderthals, while the parietal lobe of the brain (which integrates auditory, visual, and somatosensory information and is involved in mathematical processing) may be expanded in *H. sapiens* compared to Neanderthals. Amazingly, it has become clear over the past decade or so that there was extensive interbreeding between Neanderthals and modern humans. The ancestry test 23andMe says that my wife is about twice as Neanderthal as I am. We sometimes wonder what this could mean for our relationship.

Brain Architecture

All vertebrates, including humans, have a similar organization of the nervous system, divided into the same regions (forebrain, midbrain, hindbrain, and spinal cord) and similar major subregions (retina, optic tectum, cerebellum, etc.), so this vertebrate Bauplan of the nervous system (see chapter 2) is not special to us humans. We begin to home in on what is special about the human brain plan when we consider the differences in the relative sizes of brain regions between us and our closer relatives. All mammalian brains are distinguished from those of other vertebrate animals by the expansion of the cerebral cortex. Mammals evolved from reptilian ancestors. In modern reptiles, the region of the forebrain called the "dorsal pallium" is small, but in mammals, this region enlarged and evolved into the cerebral cortex. The cerebral cortex of hedgehogs and opossums, representatives of the earliest mammals, takes up only about a fifth of the brain. In monkeys, it is much larger, making up about half the brain, but in humans, it has taken over. About three-quarters of the mass of the human brain is cerebral cortex! As the cerebral cortex enlarged in the evolution of humans, it became convoluted, folded into hills (gyri) and valleys (sulci), as well as becoming thicker, with more neurons for processing information. As it enlarged, it also became subdivided into many areas, which are associated with different functions.

In the middle of the nineteenth century, a fierce debate raged about the origins of human beings. Richard Owen was a fossil hunter and a great naturalist. He was celebrated for his discovery of a huge clade of extinct reptiles: he had discovered the dinosaurs. Owen was a fierce critic of Charles Darwin and held the view that humans did not evolve from monkey-like ancestors. Instead, he believed in an ordained direction of evolution, which meant that from the beginning, humans must have come from a unique line of evolution. Owen used the brain to convince people of his view. He pointed to the differences in the size and shape of the monkey brain and the human brain as evidence against shared ancestry. On the other side of the debate was Thomas Henry Huxley, who came to be known as Darwin's bulldog. Huxley demonstrated that many of the different regions of the human brain, including many parts of the cerebral cortex, had their counterparts in the brains of monkeys and that the brains were, in fact, remarkably similar, despite their great differences in size. Nowadays, we have much greater evidence for functional similarities in many regions of the cerebral cortex, and the evidence that we evolved from ape-like ancestors has been found in our genes as well as in fossil records.

Humans belong to the simian primates, the monkeys and the apes. Current analyses of the sizes, shapes, connections, and functions of the cerebral cortex indicate that, compared to other simians, humans have expanded several regions of the neocortex, although not all areas have expanded equally. The primary

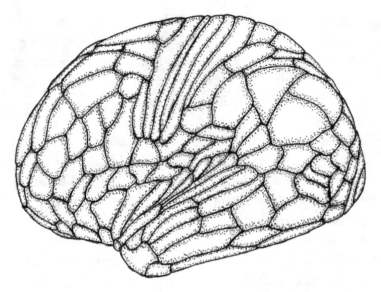

FIGURE 9.1. Multimodal analysis of the areas of the human cerebral cortex. Lateral view of left hemisphere for comparison with figure 2.4. Adapted from D. C. Van Essen, C. J. Donahue, and M. F. Glasser. 2018. "Development and Evolution of Cerebral and Cerebellar Cortex." *Brain Behav Evol* 91: 158–69, which originally appeared in M. F. Glasser, T. S. Coalson, et al. 2016. "A Multi-modal Parcellation of Human Cerebral Cortex." *Nature* 536: 171–78.

sensory and motor areas of the human cortex tend to show the least expansion. The visual cortex, for example, takes up a smaller proportion of the brain in humans than in macaque monkeys. Brodmann divided the human cortex into 52 regions (see chapter 2). Using modern techniques, such as high-resolution structural and functional magnetic resonance imaging, David van Essen at Washington University and colleagues have now identified approximately 180 different regions in each hemisphere of the human brain (figure 9.1). About 160 of these also appear in the brains of macaque monkeys, although many have expanded greatly in humans.[3] Those regions that are involved in high-level associations (e.g., areas that integrate the senses, areas that plan actions, areas that are involved in communication, and areas that are involved in abstract thinking) seem to have expanded the most. It is very intriguing that we seem to have about 20 regions that could not be identified in macaque monkeys. Are these new regions? If so, how did they arise, and what exactly do these new areas of the brain do? Given that neuroscientists are still trying to figure out how any area of the cortex, including the visual cortex, processes information—and that huge challenges arise when doing experimental work on humans—it may take a long time before any consensus emerges on what these potentially human-specific brain regions are computing. Nevertheless, it is the full combination of new cortical areas and the relative expansion and contraction of all other areas that collectively underlie the mental faculties that distinguish us from other primates.

Neoteny

In biology, the delay of a late phase of development due to the extension of an earlier phase is called "neoteny." One of the best-known examples of neoteny comes from a species of Mexican salamander, the axolotl. They are easily bred in captivity, unlike in the wild where they are threatened with extinction, and scientific laboratories throughout the world breed them and have collected various mutant lines. I used to keep a lab full of them, each with their own names—like Olivia and Newton and John. Most species of salamander go through a metamorphosis as they leave the water and become sexually mature. Adult axolotls, however, stay in the water their entire lives, maintaining their external gills and looking like larval versions of the other species.

Humans also show neoteny, which has been attributed to the brain growing faster than the rest of body. It is argued that the extensive growth of the human brain has led to a kind of limitation on pregnancy. A bigger brain means a bigger head size, which could pose a risk to the lives of mothers and babies during childbirth. We emerge from the womb immature in body as well as brain development.

Comparison of genes expressed in several regions of human and macaque monkey brains shows that both species display a marked shift in the activation patterns of thousands of genes from an early phase to a later phase.[4] This shift happens well before birth in both species. But it happens earlier in macaques than in humans, meaning that a substantial temporal delay occurs in the expression of many of the late-phase genes in humans, suggesting that the human brain stays in an early phase of growth and development for a longer period.

A recent study suggests that a gene called "ZEB2," which is associated with Mowat-Wilson syndrome (characterized by microcephaly, intellectual disability, and epilepsy), is expressed earlier in cortical organoids grown from gorilla embryonic stem cells than from human embryonic stem cells. If ZEB2 is experimentally activated early in the human organoids, the neural stem cells stop proliferating earlier and make smaller bits of cerebral cortex, or if the gene is experimentally disabled in the gorilla cells, they continue proliferating for a longer time and grow larger. These results suggest that a delay in the timing of ZEB2 activation might hold a clue to the evolution of the human brain.[5]

A functionally significant anatomical manifestation of neoteny in the human brain is that myelination is also delayed compared to other primates, such as chimps. Myelination accounts for a significant amount of postnatal brain and skull growth (see chapter 3), and so perhaps this is another reason for the delay. As a result of brain neoteny, humans are born at a stage when they must depend on caregivers for a longer time, and they have an extended period of postnatal brain refinement. Our neoteny means that experience of the world has greater impact on human brain development.

The Genes That Make a Brain Human

One-third or more of all genes we have in our genome are expressed in the developing nervous system. That is about 10,000 individual genes. When one of these becomes defective, there may be consequences for brain development or for the function or survival of neurons. Hundreds of single genes associated with different aspects of neural development, neural function, and neuronal survival have already been identified, and the list is growing rapidly with modern molecular sequencing techniques and genomic algorithms crunching the data. We have already touched on many genes that are involved in neural development in earlier chapters of this book; they are but a few of the thousands with roles in distinct aspects of neural development.

Chimps and humans have very similar genomes: 99% of the 3 billion letters of the genomes are the same. But that also means there are 30 million differences! Identifying which of these may contribute to our uniquely human brains and in which ways is a huge challenge for science. Most of the differences between the chimp and the human genome are in regions of DNA that do not make proteins. Some of these regions have evolved rapidly in the human lineage, accumulating changes at a rate that is faster than expected if there were no positive selection for these changes. It is argued that these human-specific accelerated regions (HARs) might play a role in our evolution as human beings.[6] We have clues to the functions of some HARs. For example, the very first HAR identified, HAR1, makes an RNA molecule that does not code for a protein. The exact function of this piece of RNA is still not known, but its pattern of activity is suggestive. It is active in the human fetus between weeks 7 and 18 of gestation, particularly in the reelin-secreting neurons that are involved in building the cerebral cortex (see chapter 3). Another HAR enhances Wnt signaling in the cortex (see chapter 2). The human genome also has some regions that have been duplicated only in humans, leading to second copies of some genes, which can then mutate into human-specific forms. One of these has been identified as a Notch receptor (see chapter 4) and is involved in increased proliferation. Indeed, a large fraction of the hundreds of HARs so far identified are associated with genes that have known roles in brain development.

Combining the power of genomics with mini-brain organoids has opened areas of research that seemed ridiculously fanciful when I was starting in the field. For example, a recent study focused on a gene called "Nova1," which codes for a protein involved in synapse formation. The Nova1 gene is one of a very few that make proteins containing a structural change that is shared among all humans and not shared with any of our extinct cousins, the Neanderthals of Europe and the more recently discovered Denisovans of Asia, both of which did some interbreeding with humans. In a recent set of experiments using "clustered regularly interspaced short palindromic repeats" (CRISPR)-based gene-editing techniques, the human version of the Nova1 gene was replaced in some embryonic stem cells with the archaic version of Nova1 found in Neanderthals and Denisovans.[7] The result was that the shape of cortical organoids grown from these human stem cells in culture, as well as the molecular composition and the functions of synapses that were made in these brain organoids, were slightly different from the ones found in the organoids expressing the human version of Nova1. These results suggest that the change in Nova1 may have contributed to the evolution of the human brain from our clos-est ancestors. As we are only at the beginning of such investiga-tions with mini-brains grown in tissue culture, we should inter-pret these results with great caution, yet it is certainly an exciting new avenue for research into human brain evolution.

Language

What mental functions separate us most from other animals? What makes us human? Answering this question is a continuous quest of philosophy, comparative psychology, and neuroscience. Many ideas have been put forward, including consciousness, conscience, creativity, sense of self, the ability to remember where and when events in one's life took place, a sense of fair play and morality, the ability to solve challenging problems, the ability to invent new strategies, the use of tools, and so forth. Most ideas of the distinction between animals and humans have been disputed by naturalists, ethologists, and neuroscientists, who see elephants mourning together, chimps teaching other chimps new skills that spread through the community, lions tak-ing revenge on hyenas, birds having a good idea of what was going on in the minds of other birds, animals inventing ways to get food that is otherwise out of reach (e.g., crows dropping stones into a half-filled cylinder to raise food that is floating on the water surface to an accessible level). These human observers have also witnessed macaque monkeys making economic deci-sions in a laboratory setting that are concordant with the results obtained from the sophisticated mathematical equations found in textbooks of modern economic theory for rational decision-making under various risk and reward situations.

Although arguments about what makes us human may never be fully resolved, language is generally agreed to be one of our most advanced traits as a species. But is language really a human specialty? All animals communicate. Ants leave chemical trails, bees dance to tell one another the distance and direction of nectar sources, and even bacteria signal to one another. Our closest relative, the chimp, shows very sophisticated communication skills. One chimp makes eye contact with another who is approaching and then flings his hand abruptly to the side in a gesture that means "Move away!" This is a gesture that humans also often use. Other gestural commands that seem almost intuitive to us are: "Follow me," or "Look over there!" Chimps also often accompany gestures with hooting vocalizations. Different hoots can signal alarm, patches of food, community membership, individual identity, sexual interest, and so forth. It is undeniable that chimps have a complex communication system. But language, as we humans know it, is a form of communication that employs syntax, grammatical rules, semantic constructions, and references to theoretical concepts, which can be used to package complex ideas together into a sentence that is sometimes far too long, such as this one.

If sophisticated language is particularly human, how does it feature in the human brain? In 1861, the French physician Paul Broca performed an autopsy on the brain of a 51-year-old patient, nicknamed Tan.[8] After an accident at the age of 30, Tan was able say only one word, and that one word was "Tan." He could understand language and could answer questions. For example, if you asked him to subtract 9 from 13, he would say "Tan Tan Tan Tan" to indicate 4. Broca found a lesion in Tan's brain at the back of the lateral side of the frontal lobe of the left cerebral cortex. Broca soon had another patient whose speech had been reduced to just five words. After the autopsy on the brain of the second patient, Broca wrote: "I will not deny my surprise bordering on stupefaction when I found that in my second patient the lesion was rigorously occupying the same site as the first."[9] This condition, where patients can understand spoken language but cannot speak it, is most often associated with lesions in just this part of the brain, now known as Broca's area. A decade later, another physician, Carl Wernicke, working in Austria, identified another brain area that was complementary in many ways to Broca's area.[10] Now known as Wernicke's area, when this region is damaged, patients can speak but do not understand speech or written language. When speaking, patients with damage to Wernicke's area choose the wrong words and do not make sense. For example, if asked about what they had for breakfast, they might respond: "The shoestrings under the old oak tree, singing in the sun, always so noisy, don't you know?" and think that they were answering appropriately.

Once these regions of the cortex that are involved in producing and understanding language had been identified in the human brain, it became possible to ask whether other primates also have these brain areas. The answer is yes. The structural equivalents of Broca's and Wernicke's areas have been identified in monkeys from their positions in the cortex, the specialized types of neurons they contain, their patterns of connectivity with other brain areas, and their functions in making and responding to communications. For example, stimulation of the Broca's area homologue in macaques causes mouth and face movements similar to those used in speech. Gestural communication in monkeys is associated with activity in Broca's area, and hearing species-specific calls activates both Broca's and Wernicke's areas in monkeys, just as language does in humans. So, these areas exist in nonhuman primates and are connected in ways that could pave the way for language. In humans, however, these areas have expanded, especially on the left side of the brain. The average human brain is 3.6 times larger than the chimp brain, yet the human Broca's area is almost seven times larger than it is in the chimp.

As the language areas of the brain have expanded and specialized in humans through evolution, one might predict that newborn babies already have a disposition for language. Indeed, newborns (three days old and younger) respond to recordings of spoken language more than they do to a language of nonhuman whistles. They also respond better to their native language (the one that they heard in the womb) than to a foreign language, and they respond better when native language is played forward rather than backward—so they already seem to "know" many important things about language. Electroencephalographic and functional magnetic resonance imaging studies show that areas of the left cerebral cortex, where language is encoded and decoded, light up in babies when they hear language. Remarkably, recent research has even shown that a region of the left temporal lobe known as the visual word form area (used for recognizing letters and written words) is already selectively connected to other language centers at birth.[11] Though spoken and written language may not be understood or producible for years, regions of the brain are already wired to learn to decode and produce language long before a child is able to speak.

That language-associated circuitry in the brain develops early in humans implies that genetic mechanisms are involved. This raises the possibility that one might find certain genes with key roles in the development of language in the brain. The first insight into the genetics of language came from an English family that exhibited speech defects over several generations. About half of the members of this large family showed unusual rigidity of the lower half of their faces, and most could not complete pronouncing a whole word. They would say "boon" for "spoon," "bu" for "blue." They had limited vocabulary and clearly had difficulty with producing several human speech sounds. Genetic studies of the family showed that affected individuals carried a mutation in a gene for the transcription factor FoxP2.[12] FoxP2 mutations have now been found to be involved in many individuals with similar language problems. People affected by FoxP2 mutations have measurable changes in their brains, such as gray-matter thinning in several areas of the cortex, including Broca's area. Functional imaging of the brain during language-related tasks shows that Broca's area, as well as other language-related regions, are underactive compared to unaffected relatives.

FoxP2 is not specific to humans.[13] All mammals have FoxP2. In mice, FoxP2 affects squealing. Mice with a FoxP2 mutation do not squeal as much as ordinary mice, and when they do, their squeals are abnormal. Birds also have FoxP2. In some songbirds, like zebra finches, males produce songs that they have learned from their father. FoxP2 in the brain is necessary for song learning and production. The involvement of FoxP2 in vocal communication in both mammals and birds is very suggestive that FoxP2 is important for the development of a language. But one of the things that has intrigued the scientific community most about FoxP2 is that the gene has evolved in humans. Birds, mice, and even chimpanzees all have the same form of the FoxP2 transcription factor. But the human FoxP2 has changes in a couple of the amino acids in the protein. That it is so conserved in other animals (including our closest living relatives) indicates that these changes happened relatively recently in primate evolution. DNA sequencing studies done on fragments of Neanderthal and Denisovan tissues show that these extinct hominins, with whom the *Homo sapiens* interbred, all have the form of FoxP2 that humans have, consistent with the possibility that they may also have been capable of language. We do not know how the changes in the structure of FoxP2 affect its function as a transcription factor. We do know, however, that the changes do something, as mice pups carrying a humanized version of the FoxP2 gene squeal in an unnaturally low pitch![14]

Modern genetic analysis and DNA sequencing of families and individuals with genetic disorders of speech have now identified dozens of other genes that, like FoxP2, are involved in human language.[15] Also, like FoxP2, most of these genes are not completely specific for language, as they are also associated with other cognitive syndromes, indicating that these "language genes" are also involved in other aspects of brain development. A modern view would be that building the language-specific circuitry of the brain is the task of thousands of genes working together, most of which are also involved in building other brain regions. Some fraction of these genes code for key transcription factors, like FoxP2, with regulatory roles over the expression of many other genes. It is because of the action of all these genes on developmental events—such as cell proliferation, neuronal cell type determination, axon navigation, and synapse formation—that when a baby is born, some of the baby's brain areas are enriched in the cells and the circuitry that will allow a growing child to learn to understand language, to speak, to read, and to write.

"How are you, you little pickle?" When my granddaughter was a few months old, she did not answer this question when I asked her, though she looked at me and sometimes made little cute cooing noises. Children need to hear language to understand it, and they need to practice it before they can answer such questions with what I am getting now from her at three years old: "I'm *not* picko!" It is the same for the song of the zebra finch. They learn to sing by listening to an adult song, memorizing it, and trying to reproduce it. At first, baby zebra finches babble, like human infants. They produce imperfect little chirps and fragmented pieces of song. But as the days go by, they begin to sing better and to join phrases together. They work to perfect the match between the imprinted memory of their father's song and the one that they are producing themselves. By the time a zebra finch is three months old, its song has crystalized, and it sounds very much like its father's. This is, of course, a little different from what my granddaughter is doing, which is not simply a case of imitation—she is rejecting the entire premise of my question! So clearly, something more is going on in her brain.

There is a critical period for song learning in zebra finches (see chapter 8). Do humans have a similar critical period for learning language? Frequently cited in this regard is the case of an American "feral" child named Genie born in 1957.[16] Genie's child-hating and noise-intolerant father imprisoned her in a room when she was just 20 months old. He strapped her to a toilet during the day, and he strapped her to her bed at night. She was forbidden to communicate with anyone, and she was beaten or denied food when she made any noises. He himself did not speak to her, but instead barked at her, like a dog. She was rescued by the police and hospitalized at the age of 13. Many linguists worked with her until she was 18, which is why this single case is so well documented. Genie became extremely adept at gestural communication, but her language skills improved only a very little. The implication of Genie's case is that, like a finch, a human needs to hear and produce language during early life to allow the proper refinement of the neural architecture that can fluently decode and produce language. Without input, the main language centers of the brain atrophy, as scans of Genie's brain seemed to indicate. Genie's case is one of an extremely deprived, maltreated, and malnourished child, and so cannot be used by itself to make any solid conclusions. Nevertheless, the idea that there might be a critical period for language acquisition in humans sits well with the rather undisputed fact that almost all adults have more trouble learning second languages than do almost all children and that the youngest children usually have the easiest time becoming fluent and accent-free in a second language. The decline in acquisition of proficiency in a second language begins early, for even as infants learn their native language, they begin to display a diminished ability to distinguish between speech sounds in a foreign language. For example, most American infants between six and eight months old can discriminate between two very closely related Hindi speech sounds. Yet by the time these babies are just one year old, they have already lost the ability to tell the difference. This observation also fits with studies of deaf children who were given access to hearing through cochlear implants. This work clearly indicates that implants done earlier in childhood produce more rapid development of understanding speech and producing spoken language than implants done later in childhood.

In summary, human language has been transformative for our species. It has not only allowed a sophisticated communication system between all people on the planet, but it has also served as the basis of specifically human endeavors, such as the study of history, philosophy, natural science, the spoken arts, and the passing down of the "wisdom" of the ages. Language is our way of surpassing genetics and epigenetics by recording information that is learned by our brains about the world in ways that can potentially influence billions of others now and in future generations. Language is initially fashioned in the developing human brain by thousands of genes, many of which regulate yet other genes. They work together in just the right way to build the brain circuitry that can recognize, interpret, and produce this uniquely human form of communication. The neural algorithms that are used to transform utterances into language and to decode the language we hear or read are still deeply mysterious, as are the details on how the circuitry is built. During infancy and childhood, the circuitry in this system is adjusted through real world experience of language as heard and practiced. And as the circuitry is refined and adjusted, the capacity to become fluent in second languages diminishes.

Asymmetry

The two sides of the human brain, like the brains of all other bilaterally symmetric animals, are nearly but not perfectly symmetrical.[17] They are more like slightly distorted mirror images across the midline. Whatever region is there on the right, appears to be there on the left as well, though the one on the left might be a little different from the one on the right. Detailed anatomical studies using modern imaging of cortical thickness, histology, gene-expression patterns, and connectivity patterns have shown that the human cerebral cortex is more asymmetrical than that of our closest relatives, the chimps. For example, chimps, like us, have Broca's areas on both the left and right sides, but chimps have no size asymmetry in this region, whereas humans do.

In humans, as in all vertebrates, the right brain senses what is on the left side of the body and controls the muscles on that side, and the left brain does exactly the reverse. But the human brain is not symmetrical for language. Language is much more concentrated on the left side of the brain in most people. "Nous parlons avec l'hémisphère gauche" (we speak with the left hemisphere) is what Broca said on the lateralization of language in humans. Patients with strokes in the left cerebral hemisphere lose sensation and control of their right sides and often lose the ability to communicate in language, whereas patients with strokes on the equivalent right side lose sensation and movement on the left side on the body but often retain full language capability.

The corpus callosum is an enormous tract of white matter containing about 200 million myelinated axons that connect the two hemispheres of the cerebral cortex to each other. In the 1960s, neurosurgeons looking to treat severe forms of epilepsy found that when they cut through the corpus callosum, severing all these axons, it provided great relief in some cases with what seemed like amazingly minimal disruption to brain function, considering how this huge highway of communication in the brain had been cut. The patients that received such surgery became known as "split-brain" patients. Without sophisticated tests, it is nearly impossible to tell a split-brain patient from any other person—and many people with no diagnosed neurological or cognitive problems are found, postmortem, to have been born without a corpus callosum!

Roger Sperry, who gave us the idea of chemoaffinity (see chapter 6), was also very interested in left brain vs. right brain, and the split-brain patients offered him a unique opportunity to investigate lateralization in the human brain. One immediate surprise of his work on split-brain patients was the finding that language was not totally constrained to the left side of the brain. Sperry found that there was considerable capacity for understanding language in the right hemisphere. He would flash a word on screen so that it was visible on the left side of a fixation point, so that the word was detected only by the right side of the patient's cerebral cortex. He would then ask the patient to pick up one of several objects in front of them. If, for example, he showed the word "apple" to the right side of the brain only, a split-brain patient would pick up an apple. Such tests were used to show that the right hemisphere could even understand complex phrases like "a container for liquids." Yet all the while, the speaking left hemisphere remained totally unaware of these performances of the right half, and the patient was unable to say why, for example, he had picked up the apple or the measuring cup.

In one of Sperry's lectures in the neurobiology course at Caltech, he told the students about how once he had shown a prurient image to the right hemisphere of a split-brain patient (he did not say exactly what the image was). The patient was clearly embarrassed. The left hemisphere felt that something had happened but did not know what—and when the patient was asked why he was blushing, he said he felt deeply embarrassed, but he did not know why. In split-brain patients, the two hemispheres act like separate beings; they may even have mutually conflicting cognitive operations running simultaneously, with each half inaccessible to the other hemisphere. As Sperry put it, each brain half "appeared to have its own, largely separate, cognitive domain with its own private perceptual, learning and memory experiences, all of which were seemingly oblivious of corresponding events in the other hemisphere."[18] Two semi-independent minds in one brain.

Another obvious asymmetry in the brain is handedness— whether you tend to favor the right or left hand for most tasks. By 15 weeks of gestation, most human fetuses move their right arms more than their left and seem to prefer to suck their right thumbs more than their left. Tracking the children that had these ultrasound scans showed that this asymmetric prenatal preference correlates well with handedness later. Neuroanatomical studies and gene expression studies of fetuses agree that laterality in the brain is evident prenatally.

Many people have a more even balance of handedness and language, and some people have reversed asymmetries (e.g., speech on the right side rather than on the left). Different aspects of brain asymmetry, such as language dominance and handedness, are poorly correlated with each other, so one might be left-handed but have language dominance on either the right or the left. The lateralization of handedness is much more variable than the lateralization of language—most left-handed people still speak using their left hemisphere—suggesting that different developmental mechanisms might be involved in the lateralization of different brain areas. Yet there is also an interaction because left-handers tend to have more language function on the right side of their brains. When Sperry and colleagues measured various cognitive functions in split-brain patients, he noticed that those patients who had higher scores on tasks given to the left brain showed a corresponding drop in their scores for right-hemisphere performance, and vice versa, suggesting that lateralization is variable or flexible—either side can take on a greater share than the other. Flexibility in the system is seen in children who sustain an injury to the left side of the brain at a very young age. The right cerebral cortex can take over language, and such children's language skills often become normal. Yet if a similar injury occurs in an adult, the effects on language are usually profound and irreparable.

One may speculate that the lateralization of our brains is linked to the lateralization of our guts. Almost everyone has their heart, stomach, and spleen on the left side of their body, with their liver, gallbladder, and appendix on the right, as well as a myriad of other left-right asymmetries in intestinal and vascular systems. Astonishingly, this left-right asymmetry in our bodies starts by the wriggling of a few cilia at about week three of gestation, at the time of gastrulation, when mesodermal cells begin to migrate inside the embryo at the node (see chapter 1). The synchronized beating of cilia at the node tends to push the extracellular fluid in the node from right to left. Dissolved in this continuous leftward flow is a secreted protein, appropriately called "Nodal." As a result of this leftward flow, the cells on the left of the node receive more Nodal signal than those on the right, which sets off a chain reaction of events that distributes our guts asymmetrically. A very rare condition called "situs inversus" happens in humans when genes involved in ciliary movement at the node are disrupted. Situs inversus involves complete right-to-left reversal of all the internal organs. In people with situs inversus, the relationship between the organs is not changed in any fundamental way, it is just mirror-image reversed, and most people with situs inversus have no specific medical problems. What about the brain of a person with situs inversus—is it also reversed? No, it is not! Most individuals with situs inversus have language in the left hemisphere and are right-handed. Lateralization of the brain regions therefore appears to happen independently of the lateralization of our guts.

So what do we know about the development of lateralization in the brain? First, identical twins are not much more similar to each other with respect to lateralized brain anatomy and function than nonidentical twins. This suggests that genes play very little role in the variability in lateralization for cortical regions. For language and handedness, there is instead the suggestion that exposure of the embryo to steroidal sex hormones during pregnancy, which may influence gender development, may also have some influence on lateralization. And this fits with the fact that men tend to be left-handed more frequently than women do. And women on average are a little more bilateral than men with respect to language, and so women often suffer less language breakdown when they have strokes on the left side. Lateralization remains full of mysteries: genetics seem to play a minor role, and hormonal exposure in the womb seems to play some role. Experience of the world almost certainly also plays a role. For example, the social pressure to be right-handed, which happened in the United States in the 1910s and 1920s (and continues in some places in the world), forces natural left-handers to learn to write with their right hands.

Epigenetics

Epigenetics is the study of how genes are turned on and off by means that do not involve changes to the sequence of the DNA. Changes to the sequence of DNA (i.e., genetic changes that can be inherited by the next generation) are what the modern theory of evolution is built on. Genetic changes, for example, are essential for building human brains rather than monkey brains in humans. In contrast, epigenetic changes involve the chemical linking of methyl (CH_3), acetyl (CH_3CO), and other small chemical groups to the DNA or to proteins called "histones" that pack the DNA into chromosomes. The addition of such groups can change the way genes are activated over long periods of time. For example, methyl groups can be linked directly and stably to DNA and can thus affect the expression of nearby genes through an entire lifetime. One set of epigenetic changes to the DNA and histones that occurs during development activates the genes involved in neural differentiation and represses genes that are involved in proliferation. So these changes help control the timing of brain development. Because of this, defects in making these epigenetic changes are thought to lead to some brain tumors.

It has long been known that steroidal sex hormones influence the development of gender differences in the bodies and brains of virtually all vertebrates, as well as influencing lateralization in the brain, as just mentioned. Prenatal exposure to these hormones elicits epigenetic changes in the chromosomes of developing brain cells. A pregnant mother is constantly transmitting signals to the fetus through the placenta in the form of hormones, nutrients, and other bioactive molecules. As a result, the developing embryo inside is getting some information about environmental conditions outside. For example, the various nutrients (or lack thereof) in the maternal diet; the level of oxygen; and the levels of stress, alcohol, and drugs will change the activity of enzymes that add or remove methyl and acetyl groups to the chromosomes and thus change the way genes are expressed as the brain develops and functions later. The information transferred in this epigenetic way from mother to embryo may allow a developing organism to adopt traits that better suit the environment that its mother is experiencing, but such epigenetic changes may have negative consequences if the environment changes. Most epigenetic changes are cleared from the DNA in early embryonic stages, so that the next generation starts with pristine DNA that can be epigenetically modified, but a small fraction of these changes escape erasure. Thus, one generation can potentially influence the following generations in a way that involves gene activity but does not involve changes in the sequence of the DNA.

Epigenetic changes in response to starvation have been studied in laboratory nematodes.[19] In the real world, these animals have boom-or-bust lives. They may be born on a food supply that will last through many generations of their short lives (less than three weeks), or they may be born when the food supply dries up. If one starves nematodes in the lab for one generation, it affects future generations. Under starvation conditions, a population of nematodes arises that is tougher and lives longer than the nematodes in times of plenty. This state is maintained across generations by epigenetic changes that are like a memory trace of the famine. These experiments with worms are reminiscent of the cruel experiments done on Dutch prisoners during World War II to see how people would survive on one-third of the usual caloric intake for humans. After the war, scientists were able to study the children who were born during the enforced famine. They were smaller than average, even though they had not been starved postnatally. In addition, the children of these children were also smaller. Detailed records from human populations exposed to this and other cases of undernutrition indicate that maternal starvation during pregnancy also correlates with an increased risk of adult cardiovascular and metabolic diseases, such as diabetes, in the offspring.[20]

Among laboratory rats, there are some mothers whose nature it is to be very attentive to their newborns, licking and grooming them almost constantly throughout the entire suckling period. We call them "high-lickers." Then, there are other mother rats, "low-lickers," who are much less attentive to their young. Adult female rats raised by low-licking mothers tend to become low-lickers themselves, while rat mothers who are high-lickers have generally been raised by high-licking mothers. This cycle can be broken by cross-fostering. For example, if a newborn female rat pup that is born to a low-licking mother is placed into a litter of newborn pups of a high-licking mother, she will become a high-licker herself and pass that trait on to her daughters. But such cross-fostering rarely happens in the real world, so the early effects of maternal attention (or lack thereof) promotes the continuation of these traits across generations.[21]

Studies of changes in DNA methylation in neonatal rats have shown that hundreds of genes are differentially regulated by licking and grooming. Many of these genes influence brain development. Some of the behavioral changes in rats raised by low-licking or high-licking mothers are associated with changes in the expression of genes involved in stress hormone levels in the brain. Stressful environmental conditions cause mother rats to spend less time licking and grooming their pups, which is stressful for the pups, causing their stress circuits to become overreactive and leading these circuits to become stronger. Thus, rats raised by mothers who spend little time licking and grooming them tend to become easily stressed as adults and are more fearful than are rats raised by attentive mothers.

Adult humans who have been raised with minimal attention also tend to score higher on indexes of fearfulness, negative emotions, and social inhibition, suggesting long-lasting and possibly epigenetic changes in brain development. Extreme deprivation of maternal care, such as is experienced by newborns raised in some institutionalized nurseries, has also been associated with changes in brain morphology that are not normalized by subsequent years of environmental enrichment, though sensitive and responsive caregiving from a childcare provider can ameliorate some emotional effects of such early deprivation.

Traumatic experiences in adults may also leave their imprints on the epigenome. In laboratory rats, fear conditioning (an electric shock paired with an otherwise nonfrightening stimulus) leads to changes in the methylation of hundreds of genes and therefore causes alterations in their expression. Studies in mice have even indicated that fear conditioning can be inherited by the next two generations through epigenetic mechanisms. For example, coupling an odor to an electric shock in a parent will bias the offspring of that mouse to be fearful of that odor, even though they have never smelled it before.[22] In our house, we often debate arachnophobia (fear of spiders), especially large, hairy, fast-moving spiders. My daughter freaked out when she unexpectedly came upon one while she was holding her daughter, who was two years old at the time. My granddaughter is now terrified of big hairy spiders. When I asked my daughter how she developed the fear, she could not remember, but my wife also screams when she sees a big spider, so I suspect that maybe she learned it from her mother. But my wife claims that no one taught her. She says that she was instinctively afraid. In fact, she argues that it makes good sense to be afraid of snakes and spiders, and that if it was not already programmed into us, either genetically or epigenetically, it should have been!

Variation

Our brains are identifiable as uniquely human through their common anatomical features, yet there is a tremendous amount of variability among human brains.[23] Measures of several anatomical features of the cerebral cortex show that human brains are more variable in terms of the shape, size, and thickness of different cortical regions than are those of chimpanzees. Brain size alone in humans can vary up to twofold, and region size commonly varies by this amount. The smallest variability in brain anatomy is found between identical twins. Yet MRI studies of identical twin babies show that differences in patterns of cortical folding can be used to tell them apart with 100% accuracy, just as the variations in their fingerprints can be used to tell them apart.[24] That the brains of identical twins are more alike than those of nonidentical twins speaks to inherited components that account for the variability of brain anatomy, but the many visible brain variations between the brains of identical twins speaks to acquired or chance components of brain anatomy.

Imagine scrunching up a large pizza, which is in fact about the size of an unfolded cortex, to fit inside a human skull. The cerebral cortex scrunches itself up in a coordinated way as it grows. As a result, some of the gyri and sulci are identifiable in almost everyone and are consistent enough from individual to individual to have earned names. For example, the central sulcus, which sits near the dividing line between the primary motor and primary somatosensory cortex, separates the frontal lobe from the parietal lobe. But the pizza-scrunching program is not perfectly rigid, so sulci and gyri vary in depth, length, and exact course, and there are many other smaller sulci and gyri along convolutions of the cortex that are so inconsistent that they do not have names. It is notable that the parts of the brain where the convolutions are most similar among people (e.g., the central sulcus) are closest to the parts of the human cerebral cortex that are most similar in size and shape to those in other primates, and therefore, presumably, the most evolutionarily ancient regions of the cerebral cortex. In contrast, the regions of more variation in the convolutions are found in the higher-level association areas of the cortex (i.e., the regions that have expanded and evolved rapidly in humans). The greater variation in the parts of the cerebral cortex that are newest and are concerned with this higher-level processing may be due to the shorter evolutionary history of these areas compared to the more ancient primary sensory and motor areas.

Our faces change as we age, and so do our brains. Artists and computer programs have learned to make reasonable predictions about how a child of 10 might look at the age of 40. Similar predictions can be made about the way a brain changes over the years. Multimodal MRI studies suggest that between the ages of 3 and 20, many predictable changes do indeed occur in the brain. For example, the surface area of the cerebral cortex, measured as if it were unscrunched and flattened, tends to increase between 3 and 11 years of age, and then decreases from adolescence into young adulthood. The higher association areas of the cerebral cortex show the greatest postnatal expansion, and different subparts of the cortex consistently show different trajectories from one another. The changes in brain structure between the ages of 3 and 40 years are relatively large, so it is easy to tell a toddlers' brain from an adult brain. But because many of these changes are relatively predictable, we can take these trends into account when measuring variability. If one does so, the variation between individual adult human brains is much larger than the variation between the brains of the same individuals and their predicted brains based on MRI scans as toddlers.[25] This observation suggests that experience in the postnatal world does not greatly influence the basic neuroanatomical structure of the cerebral cortex.

We do not know why human brains are so variable. Perhaps it is a matter of luck. Chance permeates development, from the flickering on and off of genes to whether or not a neural stem cell divides again (see chapter 3). Some of the differences between the brains of identical twins may be because the coordinated expression of genes is a little sloppy. Other differences between the brains of identical twins may result from randomly acquired differences in the genomes. Identical twins have thousands of small sequence differences in their DNA, which arise through replication errors and mutation events. These changes may affect the expression or the function of many genes. The high heritability of many neurological syndromes means that when one twin has such a syndrome, the other twin usually also suffers. Yet, one identical twin may suffer from such a condition, and the other twin may not. Geneticists can use these discordant identical twins to hunt through their genomes in hopes of finding those rare genes that happen to be different between the twins and thus possibly involved in neurological diseases like schizophrenia, autism, and bipolar disorders, as well as a variety of somatic diseases.

Personality and the Human Brain

Personality and character are variable among humans. Some people are more shy than others, some are more obstinate, some are more empathetic, some are more socially challenged, some are cautious, while others seem happy to take big risks, and so on. A person's full personality is constructed out of combinations of many such attributes. In adults, personality traits are rather stable from one year to the next, though brain injuries and neural degenerative diseases can change adult personality profoundly and abruptly. Personality seems to change more quickly in children and teenagers. Yet longitudinal studies show that as people reach adulthood, they generally trend in similar directions. For example, most people gradually become more agreeable and more emotionally stable than they were as youngsters.

To study personality, many British and American psychologists use scores on five traits, identified by multifactorial analyses. These scores account for quite a lot of the variation in personality types that the tests are meant to assess. As the scores involve answering questions in language, such tests cannot be administered to infants. So, how can one assess whether newborn infants already have personalities? Babies have what psychologists call "temperaments," such as whether the child is easily distracted or more stubborn, whether they are impulsive or restrained, and whether they are fearful or fearless. It is thought that elements of temperament can be regarded as precursors to the big five personality traits. For example, longitudinal studies on toddlers who were scored on a scale from inhibited to uninhibited (which considers scores on cautiousness, fearfulness, and avoidance of the unfamiliar) showed that the more "inhibited" infants were more likely to become more introverted as adults.[26] A recent study that measured activity patterns in one-month-old babies found a correlation between these patterns and aspects of temperament.[27]

The nursery rhyme says that Wednesday's child is full of woe and Friday's child is loving and giving. But the early expression of temperament and the fact that identical twins, even those separated at birth, tend to have more similar personalities than nonidentical twins when tested as adults, suggest that at least some aspects of personality have a genetic basis. Yet genetics does not account for the disparities in the personalities of identical twins, which remains one of the biggest differences between them.[28] They look more like each other than they act. Toddler temperament is only a rough indicator, and not a strict determinant, of personality—many of the most agreeable adults were the stubbornest of infants. Heritability accounts for less than half of the variability seen in the big five personality traits. Genome studies that correlate genetic changes with personality traits have identified hundreds of genetic variants that correlate with personality traits, though none appears to be strongly influential. If genes account for some of the variability in personalities among people, what accounts for the rest? We are far away from a detailed understanding of the components of personality and how they are instantiated in our brains, but nonhereditary mechanisms are certainly involved.

Experience and Deprivation

Chapter 8 visited this topic in the context of the visual deprivation experiments of Hubel and Wiesel and their effects on the
binocularity of the visual cortex during a critical period of early
postnatal life. Yet even a practiced neuroanatomist could not
easily distinguish the brain of a congenitally blind person from
one with normal sight. This absence of macroscopic structural
differences clearly does not mean that there are no differences
in the brain's microscopic structure or functional connectivity.
The same areas of the brain can look similar but function in
radically different ways in different people.[29] For example, the
visual cortexes of congenitally blind persons become active
when they read braille with their fingers. This does not happen
in sighted people. The visual cortexes of blind persons are also
active when they hear sounds and language, whereas those of
sighted people are hardly activated, if at all, by either auditory
or somatosensory inputs. Consistent with these observations
are the results of experiments using a technique called "tran-
scranial magnetic stimulation," which involves wearing a helmet
with a couple of well-positioned electromagnets that can be
briefly turned on. This magnetic pulse transiently disrupts neural activity in selected cortical regions in humans, and when it
is applied to the visual cortexes of blind individuals, it disturbs
their flow of conversation, whereas it has no such effect on
sighted people. Studies of the brains of blind children show that
their visual cortexes respond well to language as early as the age
of four years. It seems probable that the visual cortex responds
to sound and to touch even at birth, and that over the first few
years of life, active visual input eliminates or suppresses these
nonvisual inputs in sighted children, whereas they are maintained in children who are blind.

The changes in the visual cortex seen in congenitally blind people stand in contrast to what happens in individuals who become blind in adulthood. The latter fail to show responses to spoken language in their "visual" cortex, even after decades of blindness. Also, when sight is restored to people who were blind as babies, they do not see as well as those whose sight is restored after adult-onset blindness, suggesting that a sensitive period exists during early life when the visual cortex becomes more fixed for other uses in blind people, and it thereby also becomes less able to process and transmit useful visual information. The conclusion one might draw from such findings is that the visual cortex need not be "visual"; it can process other useful information that comes its way. Indeed, this may be true for every area of the cortex. In fact, it may make sense to consider the different regions of the cortex as different processing areas rather than fixed-function or modality-based areas.

There is, however, no denying that Brodmann's Area 17 (aka the primary visual cortex, or V1) receives an enormous amount of axonal input from neurons that normally carry visual information. Most of the output from the retina feeds the pathways to V1. In blind people, this input is not activated by scenes from the world, and so it carries no visual information. Although the visual cortex is being used for other tasks, this lack of visual experience has a dramatic effect on the microscopic anatomy of V1, decreasing the thickness of its gray matter. V1 thins rapidly in the blind during childhood and does not change significantly thereafter. As one might expect, this thinning does not occur in the visual cortex of an individual with adult-onset blindness. In compensation, however, the congenitally blind tend to have other areas of the cerebral cortex that are thicker than those of sighted people, and there are numerous differences in the

functional connectivity among areas of the cerebral cortex between congenitally blind and sighted people.

Marian Diamond (1926–2017) worked at the University of
California, Berkeley, and was particularly fascinated by the
question of how postnatal experience affects the microscopic
structure of the cerebral cortex. She said she was first attracted
to this question by a story she was told by Donald Hebb (see
chapter 8). It might have gone something like this:

> One day I brought home a pair of young rats as pets for my
> kids. They were named Willy and Jonathan (the rats, not the
> kids). The kids adored these rats, and the rats liked my kids.
> Willy and Jonathan had the run of the house, and the kids
> played with them all the time. I thought that these lucky pet
> rats have had the benefit of an enriched and interesting life
> compared to their poor sibs that I have back at the lab, who
> still lived in cages. So, I decided to challenge my children's rats
> against their lab-reared siblings in a maze-running competi-
> tion. Which ones would learn the way to the food the quick-
> est? You guessed it. The lab rats turned out to be no match at
> all for the pet rats.

When Diamond heard this story and talked to her colleagues,
they decided to try to repeat Hebb's rat experiment using a controlled scientific approach. They put young rats into two kinds
of cages: "enrichment cages" that were filled with toys and
housed a colony of 12 rats, and "impoverished cages" that housed
no other rats and had no toys. After a few months, the rats were
tested on lab mazes, and just as Hebb's story about pet rats
would have predicted, the enriched rats did much better. Diamond then examined the brains of the enriched and the impoverished rats under a microscope. She found a noticeable decrease

in the thickness of many areas in the cerebral cortex in the impoverished rats.[30] The decrease was only 6% on average, but it was consistent across animals over large areas of the cortex. Under the microscope, the cortexes of enriched rats could be seen to have bigger neurons with longer dendrites and an increased number of synapses, and there were more glial cells. Environmental enrichment also increased the vascularization of the cortex, aiding the delivery of oxygen and nutrients to the neurons. Diamond showed that changes in cortical structure could be found by exposing rats of any age to an enriched environment for several days, but the greatest effects were seen if the rats were exposed when they were between 60 to 90 days old. This is a period of normal development when synapse loss dominates synapse formation. The enriched environment is therefore thought to counteract this downward trend by increasing the stabilization of new and active synapses.

Recent research into the effects of environmental enrichment on the structure and function of the brain suggests that environmental enrichment leads to a greater resilience of the brain to the effects of damage, age, and certain dementias. The underlying mechanisms of deprivation are likely to involve strategies of synaptic weakening and elimination, while those of enrichment also involve strategies of synaptic strengthening and maintenance. The idea is that more enriched experience saves and enriches the connections among neurons in regions of the cortex that are involved in these experiences. One might therefore predict that some of the differences in cortical thickness that are seen in the brains of identical twins will be down to their different experiences in the world. For example, MRI imaging of the brains of identical twins both brought up in the same house, one twin of whom kept practicing the piano and became very competent, while the other stopped practicing as a child. Though the sample size is not huge, the study revealed that the musically active twins had greater cortical thickness in auditory-motor areas of the cerebral cortex.[31] While nonuse can lead to a relative reduction in the thickness of certain cortical areas, it is not clear whether enrichment and practice build up brain areas or just prevent them from shrinking. Nor is it known what the time course or sensitive periods are for most areas of the cortex. These will be challenges for future developmental neuroscientists.

In summary, many similarities and many differences in mental functions exist between us and other species because our brains are both similar and different from theirs in various ways. The brains of all vertebrates are built from the same kinds of cells and have the same Bauplan, yet the relative expansions, subdivisions, and shrinkages of different regions of the brain over evolutionary time have produced as many distinctive types of vertebrate brains as there are species of vertebrate animals. The most uniquely human feature of our brain is the expanded size and thickness of the cerebral cortex, which has become, at least anatomically, the predominant neural structure of the human brain. It is highly regionalized with specialized functional circuitry. It is where high-level information processing happens. Language, for example, the specialized form of communication that humans have, is associated with cortical regions that are particularly enlarged in human brains compared to the corresponding regions in our closest primate relatives. Another feature of the human brain is the remarkable person-to-person variability in cortical lateralization, folding patterns, area sizes, and gray-matter thickness in different regions of the cortex. Some of this variability in the brain—which correlates with differences in personality, cognitive functions, and susceptibility to various neurological and psychiatric syndromes—is accounted for by genetic differences. Other factors, including conditions in the womb, randomizing influences, and early nurturing and childhood experiences, are also involved in shaping the brain. Finally, beyond childhood, the brain continues to change and update itself through the modification of its synapses. The evolutionary history of the human brain is written in the genome, but the uniqueness of each mind is written in its unique and ever changing synaptic circuitry. It seems that what makes us all human is also what makes us all different. Everyone's body and everyone's brain differs from everyone else's from the moment of birth, and these differences grow as the final shaping of our bodies and brains is done outside the womb and incorporates our individual experiences in the world. As humans, we can exert control over the environment that provides this experience, so I can leave you with the sobering and perhaps comforting thought that we have some agency in forging the structure, function, and health of the organ that is most critical to our individual identities as human beings.

ACKNOWLEDGMENTS

I AM truly grateful for the support of Dan Sanes, Tom Reh, and Matthias Landgraf, with whom I coauthored several editions of the textbook *Development of the Nervous System,* which served as a starting point for the present effort. Many friends have offered useful advice on drafts of this book: Thank you, Marigold Acland, David Bainbridge, Michael Bate, John Bixby, Jovana Drinjakovic Fraser, Patricia Fara, Daniel Field, Fred Harris, Bob Goldstein, Jeff Harris, Simon Kerss, Chris Kintner, Robert Klepka, Gilles Laurent, Joseph Marshall, Josh Sanes, Dawn Scott, Paul Sniderman, Michael Stryker, and Gunter Wagner. Thanks also to my 2020–2021 Part 1B supervisors at Clare College, Cambridge University, and to my wife Christine Holt. Only I can be blamed for the remaining problems.

FURTHER READING

MOST OF the information in this book can be found in standard textbooks (see below), where almost all of the scientific findings are referenced. Because of space limitations, I have provided only a few references, mostly to classic papers and recent reviews in the notes to the book.

Textbooks on Brain Development

L. Bianchi. 2017. *Developmental Neurobiology*. New York: Garland Science.

M. Breedlove. 2017. *Foundations of Neural Development*. Sunderland, MA: Sinauer Associates.

S. E. Fahrbach. 2013. *Developmental Neuroscience: A Concise Introduction*. Princeton, NJ: Princeton University Press.

D. J. Price, A. P. Jarman, J. Mason, and P. Kind. 2017. *Building Brains: An Introduction to Neural Development* (2nd ed.). New York: Wiley.

D. Purves and J. Lichtman. 1984. *Principles of Neural Development*. Sunderland, MA: Sinauer Associates.

M. S. Rao and M. Jacobson (eds.). 2005. *Developmental Neurobiology* (4th ed.). New York: Springer.

D. Sanes, T. Reh, W. A. Harris, and M. Landgraf. 2019. *Development of the Nervous System* (4th ed.). Cambridge, MA: Academic Press.

Textbooks on Brain Evolution

G. Schneider. 2014. *Brain Structure and Its Origins: In Development and in Evolution of Behavior and the Mind*. Cambridge, MA: MIT Press.

G. Streider. 2004. *Principles of Brain Evolution* (4th ed.). Sunderland, MA: Sinauer Associates.

Textbooks on Developmental Biology

M. Barresi and S. Gilbert. 2020. *Developmental Biology* (12th ed.). Oxford: Oxford University Press.

G. Schoenwolf, S. Bleyl, P. Brauer, and P. Francis-West. 2021. *Larsen's Human Embryology* (6th ed.). Amsterdam: Elsevier.

L. Wolpert, C. Tickle, and A. M. Arias. 2019. *Principles of Development* (6th ed.). Oxford: Oxford University Press.

Textbooks on Neuroscience

M. Bear, B. Connors, and M. Paradiso. 2020. *Neuroscience: Exploring the Brain* (4th ed.). Burlington, MA: Jones and Bartlett.

E. Kandel, J. Koester, S. Mack, and S. Siegelbaum. 2021. *Principles of Neural Science* (6th ed.). New York: McGraw-Hill Education.

L. Luo. 2015. *Principles of Neurobiology.* New York: Garland Science.

D. Purves, G. Augustine, D. Fitzpatrick, W. Hall, A. LaMantia, L. White, R. Mooney, and M. Platt (eds.). 2018. *Neuroscience* (6th ed.). Oxford: Oxford University Press.

L. Squire, D. Berg, F. Bloom, S. du Lac, A. Ghosh, and N. C. Spitzer (eds.). 2012. *Fundamental Neuroscience* (4th ed.). Cambridge, MA: Academic Press.

Popular Books

K. Mitchell. 2018. *Innate: How the Wiring of Our Brains Shapes Who We Are.* Princeton, NJ: Princeton University Press.

S. Pinker. 2003. *The Blank Slate: The Modern Denial of Human Nature.* London: Penguin.

NOTES

Preface

1. E. Dickinson. 1960. *The Complete Poems of Emily Dickinson*. T. H. Johnson (ed.). Boston: Little, Brown & Co. Part 1, Life number 126.
2. S. B. Carroll. 2011. *Endless Forms Most Beautiful: The New Science of Evo Devo and the Making of the Animal Kingdom*. London: Quercus.

Chapter 1

1. W. Roux. 1888. "Beiträge zur Entwickelungsmechanik des Embryo. Über die künstliche Hervorbringung halber Embryonen durch Zerstörung einer der beiden ersten Furchungskugeln, sowie über die Nachentwickelung (Postgeneration) der fehlenden Körperhälfte." *Virchows Arch Pathol Anat Physiol Klin Med* 114: 113–53. Translated in B. Whittier and J. M. Oppenheimer (eds.). 1974. *Foundations of Experimental Embryology*. New York: Hafner Press, pp. 2–37.
2. H. Driesch. 1891. "Entwicklungsmechanische Studien: I. Der Werthe der beiden ersten Furchungszellen in der Echinogdermenentwicklung. Experimentelle Erzeugung von Theil- und Doppelbildungen. II. Über die Beziehungen des Lichtez zur ersten Etappe der thierischen Form-bildung." *Zeitschrift für wissenschaftliche Zoologie* 53: 160–84. Translated in B. H. Willier and J. M. Oppenheimer (eds.). 1974. "The Potency of the First Two Cleavage Cells in Echinoderm Development. Experimental Production of Partial and Double Formations." In *Foundations of Experimental Embryology*. New York: McMillan, pp. 38–50.
3. H. Spemann. 1903. "Entwickelungsphysiologische Studien am Tritonei III." *Arch f Entw Mech* 16: 551–631. H. Spemann. 1938. *Embryonic Development and Induction*. New Haven, CT: Yale University Press.
4. A. K. Tarakowski. 1959. "Experiments on the Development of Isolated Blastomeres of Mouse Eggs." *Nature* 184: 1286–87.

5. W. B. Kristan, Jr. 2016. "Early Evolution of Neurons." *Curr Biol* 26: R949–R954. D. Arendt. 2021. "Elementary Nervous Systems." *Phil Trans R Soc Lond B Biol Sci* 376: 20200347. M. G. Paulin and J. Cahill-Lane. 2021. "Events in Early Nervous System Evolution." *Top Cogn Sci* 13: 25–44.

6. S. M. Suryanarayana, B. Robertson, P. Wallén, and S. Grillner. 2017. "The Lamprey Pallium Provides a Blueprint of the Mammalian Layered Cortex." *Curr Biol* 27: 3264–77.

7. H. Spemann. 1918. "Über die Determination der ersten Organanlagen des Amphibienembryo I–IV." *Arch f Entwicklungsmech d Organismen* 43: 448–555.

8. See H. Spemann and H. Mangold. 1924. "Induction of Embryonic Primordia by Implantation of Organizers from a Different Species." Translated in *J Dev Biol* 45: 13–38. (2001).

9. H. Spemann. 1935. Nobel Prize lecture. https://www.nobelprize.org/prizes /medicine/1935/spemann/lecture/.

10. J. Holtfreter and V. Hamburger. 1955. "Amphibians." In B. H. Willier, P. Weiss, and V. Hamburger (eds.). *Analysis of Development.* Philadelphia: Saunders, pp. 230–396.

11. W. C. Smith and R. M. Harland. 1992. "Expression Cloning of Noggin, a New Dorsalizing Factor Localized to the Spemann Organizer in *Xenopus* Embryos." *Cell* 70: 829–40.

12. A. Hemmati-Brivanlou and D. A. Melton. 1994. "Inhibition of Activin Receptor Signaling Promotes Neuralization in *Xenopus.*" *Cell* 77: 273–81.

13. M. Z. Ozair, C. Kintner, and A. Hemmati-Brivanlou. 2013. "Neural Induction and Early Patterning in Vertebrates." *Wiley Interdiscip Rev Dev Biol* 2: 479–98.

14. V. François and E. Bier. 1995. "*Xenopus* Chordin and *Drosophila* Short Gastrulation Genes Encode Homologous Proteins Functioning in Dorsal-Ventral Axis Formation." *Cell* 80: 19–20.

15. J. B. Gurdon, T. R. Elsdale, and M. Fischberg. 1958. "Sexually Mature Individuals of *Xenopus laevis* from the Transplantation of Single Somatic Nuclei." *Nature* 182: 64–65.

16. K. Eggan, K. Baldwin, M. Tackett, J. Osborne, J. Gogos, A. Chess, R. Axel, and R. Jaenisch. 2004. "Mice Cloned from Olfactory Sensory Neurons." *Nature* 428: 44–49.

17. M. Eiraku, N. Takata, H. Ishibashi, T. Adachi, and Y. Sasai. 2011. "Self-Organizing Optic-Cup Morphogenesis in Three-Dimensional Culture." *Nature* 472: 51–58. K. Muguruma and Y. Sasai. 2012. "In Vitro Recapitulation of Neural Development Using Embryonic Stem Cells: From Neurogenesis to Histogenesis." *Development Growth and Differentiation* 54: 349–57.

18. M. A. Lancaster, N. S. Corsini, S. Wolfinger, E. H. Gustafson, A. W. Phillips, T. R. Burkard, T. Otani, F. J. Livesey, and J. A. Knoblich. 2017. "Guided Self-

Organization and Cortical Plate Formation in Human Brain Organoids." *Nat Biotech* 35: 659–66.

Chapter 2

1. See https://www.uclh.nhs.uk/patients-and-visitors/patient-information-pages /management-fetal-spina-bifida.

2. S. J. Gould. 1977. *Ontogeny and Phylogeny*. Cambridge, MA: Harvard University Press.

3. R. P. Elinson. 1987. "Changes in Developmental Patterns: Embryos of Amphibians with Large Eggs." In R. A. Raff and E. C. Raff (eds.). *Development as an Evolutionary Process*. New York: Liss, pp. 1–21. D. Duboule. 1994. "Temporal Colinearity and the Phylotypic Progression: A Basis for the Stability of a Vertebrate Bauplan and the Evolution of Morphologies through Heterochrony." *Dev Suppl* 1994: 135–42.

4. N. Holmgren. 1925. "Points of View Concerning Forebrain Morphology in Higher Vertebrates." *Acta Zool Stochholm* 6: 413–77.

5. J. Kaas and C. Collins. 2001. "Variability in the Sizes of Brain Parts." *Behav Brain Sci* 24: 288–90.

6. M. McKeown, S. L. Brusatte, T. E. Williamson, J. A. Schwab, T. D. Carr, I. B. Butler, A. Muir, et al. 2020. "Neurosensory and Sinus Evolution as Tyrannosauroid Dinosaurs Developed Giant Size: Insight from the Endocranial Anatomy of *Bistahieversor sealeyi*." *Anat Rec* 303: 1043–59.

7. E. B. Lewis. 1957. "Leukemia and Ionizing Radiation." *Science* 125: 965–72.

8. E. B. Lewis. 1978. "A Gene Complex Controlling Segmentation in *Drosophila*." *Nature* 276: 565–70.

9. J. J. Stuart, S. J. Brown, R. W. Beeman, and R. E. Denell. 1991. "A Deficiency of the Homeotic Complex of the Beetle Tribolium." *Nature* 350: 72–74.

10. T. A. Tischfield, T. M. Bosley, M. A. Salih, A. I. Alorainy, E. C. Sener, M. J. Nester, D. T. Oystreck, et al. 2005. "Homozygous HOXA1 Mutations Disrupt Human Brainstem, Inner Ear, Cardiovascular and Cognitive Development." *Nat Genet* 37: 1035–37.

11. P. D. Nieuwkoop. 1952. "Activation and Organization of the Central Nervous System in Amphibians. Part III. Synthesis of a New Working Hypothesis." *J Exp Zool* 120: 83–108.

12. C. Nolte, B. De Kumar, and R. Krumlauf. 2019. "Hox Genes: Downstream 'Effectors' of Retinoic Acid Signaling in Vertebrate Embryogenesis." *Genesis* 57: 7–8.

13. E. Wieschaus, C. Nüsslein-Volhard, and G. Jürgens. 1984. "Mutations Affecting the Pattern of the Larval Cuticle in *Drosophila melanogaster*: III. Zygotic Loci on the X-Chromosome and Fourth Chromosome." *Wilhelm Roux Arch Dev Biol* 193:

296–307. G. Jürgens, E. Wieschaus, C. Nüsslein-Volhard, and H. Kluding. 1984. "Mutations Affecting the Pattern of the Larval Cuticle in *Drosophila melanogaster:* II. Zygotic Loci on the Third Chromosome." *Wilhelm Roux Arch Dev Biol* 193: 283–95. C. Nüsslein-Volhard, E. Wieschaus, and H. Kluding. 1984. "Mutations Affecting the Pattern of the Larval Cuticle in *Drosophila melanogaster:* I. Zygotic Loci on the Second Chromosome." *Wilhelm Roux Arch Dev Biol* 193: 267–82.

14. J. Briscoe, A. Pierani, T. M. Jessell, and J. A. Ericson. 2000. "A Homeodomain Protein Code Specifies Progenitor Cell Identity and Neuronal Fate in the Ventral Neural Tube." *Cell* 101: 435–45.

15. L. J. Wolpert. 1969. "Positional Information and the Spatial Pattern of Cellular Differentiation." *Theor Biol* 25: 1–47.

16. R. Nusse, A. van Ooyen, D. Cox, Y. K. Fung, and H. Varmus. 1984. "Mode of Proviral Activation of a Putative Mammary Oncogene (int-1) on Mouse Chromosome 15." *Nature* 307: 131–36.

17. D. Arendt, A. S. Denes, G. Jékely, and K. Tessmar-Raible. 2008. "The Evolution of Nervous System Centralization." *Philos Trans R Soc Lond B Biol Sci* 363: 1523–28.

18. M. K. Cooper, J. A. Porter, K. E. Young, and P. A. Beachy. 1998. "Teratogen-Mediated Inhibition of Target Tissue Response to Shh Signaling." *Science* 280: 1603–7.

19. W. J. Gehring. 1996. "The Master Control Gene for Morphogenesis and Evolution of the Eye." *Genes Cells* 1: 11–15. W. J. Gehring. 2014. "The Evolution of Vision." *Interdiscip Rev Dev Biol* 3: 1–40.

20. M. E. Zuber, G. Gastri, A. S. Viczian, G. Barsacchi, and W. A. Harris. 2003. "Specification of the Vertebrate Eye by a Network of Eye Field Transcription Factors." *Development* 130: 5155–67.

21. K. Brodmann. 1909. *Vergleichende Lokalisationslehre der Großhirnrinde.* Leipzig: Verlag von Johanne Ambrosius Barth.

Chapter 3

1. M. Florio and W. B. Huttner. 2014. "Neural Progenitors, Neurogenesis and the Evolution of the Neocortex." *Development* 141: 2182–94. J. H. Lui, D. V. Hansen, and A. R. Kriegstein. 2011. "Development and Evolution of the Human Neocortex." *Cell* 146: 18–36.

2. T. Otani, M. C. Marchetto, F. H. Gage, B. D. Simons, and F. J. Livesey. 2016. "2D and 3D Stem Cell Models of Primate Cortical Development Identify Species-Specific Differences in Progenitor Behavior Contributing to Brain Size." *Cell Stem Cell* 18: 467–80.

3. V. C. Twitty. 1966. *Of Scientists and Salamanders.* San Francisco: W. H. Freeman.

4. J. E. Sulston, E. Schierenberg, J. G. White, and J. N. Thomson. 1983. "The Embryonic Cell Lineage of the Nematode *Caenorhabditis elegans.*" *Dev Biol* 100: 64–119.

J. E. Sulston and H. R. Horvitz. 1977. "Post-embryonic Cell Lineages of the Nematode, *Caenorhabditis elegans.*" *Dev Biol* 56: 110–56.

5. J. He, G. Zhang, A. D. Almeida, M. Cayouette, B. D. Simons, and W. A. Harris. 2012. "How Variable Clones Build an Invariant Retina." *Neuron* 75: 786–98.

6. D. Morgan. 2006. *The Cell Cycle: Principles of Control.* Oxford: Oxford University Press.

7. O. Warburg. 1956. "On the Origin of Cancer Cells." *Science* 123: 309–14.

8. M. E. Zuber, M. Perron, A. Philpott, A. Bang, and W. A. Harris. 1999. "Giant Eyes in *Xenopus laevis* by Overexpression of XOptx2." *Cell* 98: 341–52.

9. G. K. Thornton and C. G. Woods. 2009. "Primary Microcephaly: Do All Roads Lead to Rome?" *Trends Genet* 25: 501–10.

10. J. B. Angevine and R. L. Sidman. 1961. "Autoradiographic Study of Cell Migration during Histogenesis of Cerebral Cortex in the Mouse." *Nature* 192: 766–68.

11. D. S. Rice and T. Curran. 2001. "Role of the Reelin Signaling Pathway in Central Nervous System Development." *Annu Rev Neurosci* 24: 1005–39.

12. K. L. Spalding, R. D. Bhardwaj, B. A. Buchholz, H. Druid, and J. Frisén. 2005. "Retrospective Birth Dating of Cells in Humans." *Cell* 122: 133–43.

13. A. J. Fischer, J. L. Bosse, and H. M. El-Hodiri. 2013. "The Ciliary Marginal Zone (CMZ) in Development and Regeneration of the Vertebrate Eye." *Exp Eye Res* 116: 199–204.

14. J. Altman. 1962. "Are New Neurons Formed in the Brains of Adult Mammals?" *Science* 135: 1127–28.

15. G. Kempermann, F. G. Gage, L. Aigner, H. Song, M. A. Curtis, S. Thuret, H. G. Kuhn, et al. 2018. "Human Adult Neurogenesis: Evidence and Remaining Questions." *Cell Stem Cell* 23: 25–30.

Chapter 4

1. H. Zeng and J. R. Sanes. 2017. "Neuronal Cell-Type Classification: Challenges, Opportunities and the Path Forward." *Nat Rev Neurosci* 18: 530–46.

2. F. Jimenez and J. A. Campos-Ortega. 1979. "A Region of the *Drosophila* Genome Necessary for CNS Development." *Nature* 282: 310–12.

3. R. L. Davis, H. Weintraub, and A. B. Lassar. 1987. "Expression of a Single Transfected cDNA Converts Fibroblasts to Myoblasts." *Cell* 51: 987–1000.

4. S. Ramón y Cajal. 1989. *Recollections of My Life.* Cambridge, MA: MIT Press.

5. C. S. Sherrington. 1906. *The Integrative Action of the Nervous System.* Oxford: Oxford University Press.

6. S. Ramón y Cajal. 1995. *Histology of the Nervous System of Man and Vertebrates* (translated from French by Neely Swanson and Larry W. Swanson). Oxford: Oxford University Press.

7. Ramón y Cajal. *Recollections of My Life.*

8. J. Liu and J. R. Sanes. 2017. "Cellular and Molecular Analysis of Dendritic Morphogenesis in a Retinal Cell Type That Senses Color Contrast and Ventral Motion." *J Neurosci* 37: 12247–62.

9. See https://en.wikipedia.org/wiki/Sydney_Brenner.

10. D. F. Ready, T. E. Hanson, and S. Benzer. 1976. "Development of the *Drosophila* Retina, a Neurocrystalline Lattice." *Dev Biol* 53: 217–40.

11. T. M. Jessell. 2000. "Neuronal Specification in the Spinal Cord: Inductive Signals and Transcriptional Codes." *Nat Rev Genet* 1: 20–29.

12. N. Le Douarin. 1980. "Migration and Differentiation of Neural Crest Cells." *Curr Top Dev Biol* 16: 31–85.

13. J. I. Johnsen, C. Dyberg, and M. Wickström. 2019. "Neuroblastoma—A Neural Crest Derived Embryonal Malignancy." *Front Mol Neurosci* 12: 9.

14. C. Q. Doe. 2017. "Temporal Patterning in the *Drosophila* CNS." *Annu Rev Cell Dev Biol* 33: 219–40.

15. S. K. McConnell. 1991. "The Generation of Neuronal Diversity in the Central Nervous System." *Annu Rev Neurosci* 14: 269–300.

16. C. H. Waddington. 1956. *Principles of Embryology*. London: George Allen & Unwin.

17. P. M. Smallwood, Y. Wang, and J. Nathans. 2002. "Role of a Locus Control Region in the Mutually Exclusive Expression of Human Red and Green Cone Pigment Genes." *Proc Natl Acad Sci USA* 99: 1008–11.

18. M. Perry, M. Kinoshita, G. Saldi, L. Huo, K. Arikawa, and C. Desplan. 2016. "Molecular Logic behind the Three-Way Stochastic Choices That Expand Butterfly Colour Vision." *Nature* 535: 280–84.

19. K. Yamakawa, Y. K. Huot, M. A. Haendelt, R. Hubert, X. N. Chen, G. E. Lyons, and J. R. Korenberg. 1998. "DSCAM: A Novel Member of the Immunoglobulin Superfamily Maps in a Down Syndrome Region and Is Involved in the Development of the Nervous System." *Hum Mol Genet* 7: 227–37.

20. D. Schmucker, J. C. Clemens, H. Shu, C. A. Worby, J. Xiao, M. Muda, J. E. Dixon, and S. L. Zipursky. 2000. "*Drosophila* DSCAM Is an Axon Guidance Receptor Exhibiting Extraordinary Molecular Diversity." *Cell* 101: 671–84.

21. W. V. Chen and T. Maniatis. 2013. "Clustered Protocadherins." *Development* 140: 3297–3302.

22. X. Duan, A. Krishnaswamy, I. De la Huert, and J. R. Sanes. 2014. "Type II Cadherins Guide Assembly of a Direction-Selective Retinal Circuit." *Cell* 158(4): 793–807.

Chapter 5

1. R. Harrison. 1910. "The Outgrowth of the Nerve Fiber as a Mode of Protoplasmic Movement." *J Exp Zool* 9: 787–846.

2. R. G. Harrison. 1907. "Observations on the Living Developing Nerve Fiber." *Proc Soc Exp Biol Med* 4: 140–43.

3. S. Ramón y Cajal. 1989. *Recollections of My Life.* Cambridge, MA: MIT Press.

4. Ramón y Cajal. *Recollections of My Life.*

5. C. M. Bate. 1976. "Pioneer Neurones in an Insect Embryo." *Nature* 260: 54–56.

6. D. Bentley and M. Caudy. 1983. "Pioneer Axons Lose Directed Growth after Selective Killing of Guidepost Cells." *Nature* 304: 62–65.

7. C. S. Goodman, C. M. Bate, and N. C. Spitzer. 1981. "Embryonic Development of Identified Neurons: Origin and Transformation of the H Cell." *J Neurosci* 1: 94–102. M. Bate, C. S. Goodman, and N. C. Spitzer. 1981. "Embryonic Development of Identified Neurons: Segment-Specific Differences in the H Cell Homologues." *J Neurosci* 1: 103–6.

8. J. A. Raper, M. J. Bastiani, and C. S. Goodman. 1984. "Pathfinding by Neuronal Growth Cones in Grasshopper Embryos. IV. The Effects of Ablating the A and P Axons upon the Behavior of the G Growth Cone." *J Neurosci* 4: 2329–45. M. J. Bastiani, J. A. Raper, and C. S. Goodman. 1984. "Pathfinding by Neuronal Growth Cones in Grasshopper Embryos. III. Selective Affinity of the G Growth Cone for the P Cells within the A/P Fascicle." *J Neurosci* 4: 2311–28. J. A. Raper, M. Bastiani, and C. S. Goodman. 1983. "Pathfinding by Neuronal Growth Cones in Grasshopper Embryos. II. Selective Fasciculation onto Specific Axonal Pathways." *J Neurosci* 3: 31–41. J. A. Raper, M. Bastiani, and C. S. Goodman. 1983. "Pathfinding by Neuronal Growth Cones in Grasshopper Embryos. I. Divergent Choices Made by the Growth Cones of Sibling Neurons." *J Neurosci* 3: 20–30.

9. W. A. Harris. 1986. "Homing Behaviour of Axons in the Embryonic Vertebrate Brain." *Nature* 320: 266–69.

10. J. S. Taylor. 1990. "The Directed Growth of Retinal Axons towards Surgically Transposed Tecta in *Xenopus:* An Examination of Homing Behaviour by Retinal Ganglion Cells. *Development* 108: 147–58.

11. E. Hibbard. 1965. "Orientation and Directed Growth of Mauthner's Cell Axons from Duplicated Vestibular Nerve Roots." *Exp Neurol* 13: 289–301.

12. W. A. Harris. 1989. "Local Positional Cues in the Neuroepithelium Guide Retinal Axons in Embryonic *Xenopus* Brain." *Nature* 339: 218–21.

13. A. G. Lumsden and A. M. Davies. 1986. "Chemotropic Effect of Specific Target Epithelium in the Developing Mammalian Nervous System." *Nature* 323: 538–39.

14. E. M. Hedgecock, J. G. Culotti, and D. H. Hall. 1990. "The *unc-5, unc-6,* and *unc-40* Genes Guide Circumferential Migrations of Pioneer Axons and Mesodermal Cells on the Epidermis in *C. elegans." Neuron* 4: 61–85.

15. T. E. Kennedy, T. Serafini, J. R. de la Torre, and M. Tessier-Lavigne. 1994. "Netrins Are Diffusible Chemotropic Factors for Commissural Axons in the Embryonic Spinal Cord." *Cell* 78: 425–35.

16. A. Méneret, E. A. Franz, O. Trouillard, T. C. Oliver, Y. Zagar, S. P. Robertson, Q. Welniarz, et al. 2017. "Mutations in the Netrin-1 Gene Cause Congenital Mirror Movements." *J Clin Invest* 127: 3923–36.

17. Y. Luo, D. Raible, and J. A. Raper. 1993. "Collapsin: A Protein in Brain That Induces the Collapse and Paralysis of Neuronal Growth Cones." *Cell* 75: 217–27.

18. M. Gorla and G. J. Bashaw. 2020. "Molecular Mechanisms Regulating Axon Responsiveness at the Midline." *Dev Biol* 466: 12–21.

19. W. A. Harris, C. E. Holt, and F. Bonhoeffer. 1987. "Retinal Axons with and without Their Somata, Growing to and Arborizing in the Tectum of *Xenopus* embryos: A Time-Lapse Video Study of Single Fibres in vivo." *Development* 101: 123–33.

20. C. E. Holt, K. C. Martin, and E. M. Schuman. 2019. "Local Translation in Neurons: Visualization and Function." *Nat Struct Mol Biol* 26(7): 557–66.

21. Ramón y Cajal. *Recollections of My Life.*

22. See https://www.themiamiproject.org.

Chapter 6

1. R. W. Sperry. 1945. "The Problem of Central Nervous Reorganization after Nerve Regeneration and Muscle Transposition." *Q Rev Biol* 20: 311–69.

2. C. Lance-Jones and L. J. Landmesser. 1980. "Motoneurone Projection Patterns in the Chick Hind Limb Following Early Partial Reversals of the Spinal Cord." *J Physiol* 302: 581–602.

3. R. W. Sperry. 1943. "Effect of 180 Rotation of the Retinal Field on Visuomotor Coordination." *J Exp Zool* 92: 263–79.

4. R.W. Sperry. 1943. "Chemoaffinity in the Orderly Growth of Nerve Fiber Patterns and Connections." *Proc Natl Acad Sci USA* 50: 703–10.

5. C. E. Holt. 1984. "Does Timing of Axon Outgrowth Influence Initial Retinotectal Topography in *Xenopus? J Neurosci* 4: 1130–52.

6. J. Walter, S. Henke-Fahle, and F. Bonhoeffer. 1987. "Avoidance of Posterior Tectal Membranes by Temporal Retinal Axons." *Development* 101: 909–13. J. Walter, B. Kern-Veits, J. Huf, B. Stolze, and F. Bonhoeffer. 1987. "Recognition of Position-Specific Properties of Tectal Cell Membranes by Retinal Axons in vitro." *Development* 101: 685–96. U. Drescher, C. Kremoser, C. Handwerker, J. Löschinger, M. Noda, and

F. Bonhoeffer. 1995. "In vitro Guidance of Retinal Ganglion Cell Axons by RAGS, a 25 kDa Tectal Protein Related to Ligands for Eph Receptor Tyrosine Kinases." *Cell* 82: 359–70.

7. H. J. Cheng, M. Nakamoto, A. D. Bergemann, and J. G. Flanagan. 1995. "Complementary Gradients in Expression and Binding of ELF-1 and Mek4 in Development of the Topographic Retinotectal Projection Map." *Cell* 82: 371–81.

8. J. R. Sanes and S. L. Zipursky. 2020. "Synaptic Specificity, Recognition Molecules, and Assembly of Neural Circuits." *Cell* 181: 536–56.

9. F. Ango, G. di Cristo, H. Higashiyama, V. Bennett, P. Wu, and Z. J. Huang. 2004. "Ankyrin-Based Subcellular Gradient of Neurofascin, an Immunoglobulin Family Protein, Directs GABAergic Innervation at Purkinje Axon Initial Segment." *Cell* 119: 257–72.

10. J. R. Sanes and J. W. Lichtman. 1999. "Development of the Vertebrate Neuromuscular Junction." *Annu Rev Neurosci* 22: 389–442.

11. J. E. Vaughn, C. K. Henrikson, and J. Grieshaber. 1974. "A Quantitative Study of Synapses on Motor Neuron Dendritic Growth Cones in Developing Mouse Spinal Cord." *J Cell Biol* 60: 664–72.

12. Y. H. Takeo, S. A. Shuster, L. Jiang, M. C. Hu, D. J. Luginbuhl, T. Rülicke, X. Contreras, et al. 2021. "GluD2- and Cbln1-Mediated Competitive Interactions Shape the Dendritic Arbors of Cerebellar Purkinje Cells." *Neuron* 109: 629–44.

13. F. W. Pfrieger and B. A. Barres. 1997. "Synaptic Efficacy Enhanced by Glial Cells in Vitro." *Science* 277: 1684–87.

14. B. Barres. 2018. *The Autobiography of a Transgender Scientist.* Cambridge, MA: MIT Press.

15. W. A. Harris. 1984. "Axonal Pathfinding in the Absence of Normal Pathways and Impulse Activity." *J Neurosci* 4: 1153–62. P. R. Hiesinger, R. G. Zhai, Y. Zhou, T. W. Koh, S. Q. Mehta, K. L. Schulze, Y. Cao, et al. 2006. "Activity-Independent Prespecification of Synaptic Partners in the Visual Map of *Drosophila*." *Curr Biol* 16: 1835–43.

Chapter 7

1. R. R. Buss, W. Sun, and R. W. Oppenheim. 2006. "Adaptive Roles of Programmed Cell Death during Nervous System Development." *Annu Rev Neurosci* 29: 1–35. R. W. Oppenheim. 1991. "Cell Death during Development of the Nervous System." *Annu Rev Neurosci* 14: 453–501.

2. J. E. Sulston and H. R. Horvitz. 1977. "Post-embryonic Cell Lineages of the Nematode, *Caenorhabditis elegans*." *Dev Biol* 56: 110–56.

3. P. O. Kanold and H. J. Luhmann. 2010. "The Subplate and Early Cortical Circuits." *Annu Rev Neurosci* 33: 23–48. M. Riva, I. Genescu, C. Habermacher, D. Orduz,

F. Ledonne, F. M. Rijli, G. López-Bendito, et al. 2019. "Activity-Dependent Death of Transient Cajal-Retzius Neurons Is Required for Functional Cortical Wiring." *Elife* 31: 8.

4. V. Hamburger. 1952. "Development of the Nervous System." *Ann N Y Acad Sci* 55: 117–32.

5. R. Levi-Montalcini and G. Levi. 1944. "Correleziani nello svillugo tra varie parti del sistema nervoso. I. Consequenze della demolizione delle abbozzo di un arts sui centri nervosi nell' embrione di pollo." *Comment Pontif Acad Sci* 8: 527–68.

6. V. Hamburger and R. Levi-Montalcini. 1949. "Proliferation, differentiation and degeneration in the spinal ganglia of the chick embryo under normal and experimental conditions." *J Exp Zool* 111: 457–502.

7. M. Hollyday and V. Hamburger. 1976. "Reduction of the Naturally Occurring Motor Neuron Loss by Enlargement of the Periphery." *J Comp Neurol* 170: 311–20.

8. S. Cohen, R. Levi-Montalcini, and V. Hamburger. 1954. "A Nerve Growth-Stimulating Factor Isolated from Sarcomas 37 and 180." *Proc Natl Acad Sci USA* 40: 1014–18.

9. S. Cohen and R. Levi-Montalcini. 1956. "A Nerve Growth-Stimulating Factor Isolated from Snake Venom." *Proc Natl Acad Sci USA* 42: 571–74.

10. W. M. Cowan. 2001. "Viktor Hamburger and Rita Levi-Montalcini: The Path to the Discovery of Nerve Growth Factor." *Annu Rev Neurosci* 24: 551–600.

11. H. M. Ellis and H. R. Horvitz. 1986. "Genetic Control of Programmed Cell Death in the Nematode C. elegans." *Cell* 44: 817–29.

12. M. C. Raff. 1992. "Social Controls on Cell Survival and Cell Death." *Nature* 356: 397–400.

13. R. Levi-Montalcini. 1948. "Consequences of the Eradication of the Otocyst on the Development of the Acoustic Centers in the Chicken Embryo." *Schweiz Med Wochenschr* 78: 412.

14. I. Gonsalvez, R. Baror, P. Fried, E. Santarnecchi, and A. Pascual-Leone. 2017. "Therapeutic Noninvasive Brain Stimulation in Alzheimer's Disease." *Curr Alzheimer Res* 14: 362–76. D. S. Xu and F. A. Ponce. 2017. "Deep Brain Stimulation for Alzheimer's Disease." *Curr Alzheimer Res* 14: 356–61.

15. M. Denaxa, G. Neves, J. Burrone, and V. Pachnis. 2018. "Homeostatic Regulation of Interneuron Apoptosis during Cortical Development." *J Exp Neurosci* 5: 12. F. K. Wong and O. Marín. 2019. "Developmental Cell Death in the Cerebral Cortex." *Annu Rev Cell Dev Biol* 35: 523–42.

Chapter 8

1. S. S. Freeman, A. G. Engel, and D. B. Drachman. 1976. "Experimental Acetylcholine Blockade of the Neuromuscular Junction. Effects on End Plate and Muscle Fiber Ultrastructure." *Ann N Y Acad Sci* 274: 46–59. E. M. Callaway and D. C. Van

Essen. 1989. "Slowing of Synapse Elimination by Alpha-Bungarotoxin Superfusion of the Neonatal Rabbit Soleus Muscle." *Dev Biol* 131: 356–65.

2. D. H. Hubel and T. N. Wiesel. 1977. "Ferrier Lecture. Functional Architecture of Macaque Monkey Visual Cortex." *Proc R Soc Lond B Biol Sci* 198: 1–59.

3. E. I. Knudsen. 1999. "Mechanisms of Experience-Dependent Plasticity in the Auditory Localization Pathway of the Barn Owl." *J Comp Physiol A* 185: 305–21.

4. K. Lorenz. 1935. "Der Kumpan in der Umwelt des Vogels. Der Artgenosse als auslösendes Moment sozialer Verhaltensweisen." *Journal für Ornithologie* 83: 137–215.

5. N. T. Burley. 2006. "An Eye for Detail: Selective Sexual Imprinting in Zebra Finches." *Evolution* 60: 1076–85.

6. D. O. Hebb. 1949. *The Organization of Behavior.* New York: Wiley and Sons, p. 62.

7. M. Meister, R. O. Wong, D. A. Baylor, and C. J. Shatz. 1991. "Synchronous Bursts of Action Potentials in Ganglion Cells of the Developing Mammalian Retina." *Science* 252: 939–43.

8. C. J. Shatz. 1996. "Emergence of Order in Visual System Development." *Proc Natl Acad Sci USA* 93: 602–8.

9. L. A. Kirkby, G. S. Sack, A. Firl, and M. B. Feller. 2013. "A Role for Correlated Spontaneous Activity in the Assembly of Neural Circuits." *Neuron* 80: 1129–44.

10. L. Glück. 1996. *Meadowlands.* Hopewell, NJ: Ecco Press, p. 43.

11. J. S. Espinosa and M. P. Stryker. 2012. "Development and Plasticity of the Primary Visual Cortex." *Neuron* 75: 230–49.

12. L. I. Zhang, H. W. Tao, C. E. Holt, W. A. Harris, and M. Poo. 1998. "A Critical Window for Cooperation and Competition among Developing Retinotectal Synapses." *Nature* 395: 37–44.

Chapter 9

1. G. Streider. 2004. *Principles of Brain Evolution.* Sunderland, MA: Sinauer.

2. A. A. Polilov. 2012. "The Smallest Insects Evolve Anucleate Neurons." *Arthropod Struct Dev* 41: 29–34.

3. D. C. Van Essen, C. J. Donahue, T. S. Coalson, H. Kennedy, T. Hayashi, and M. F. Glasser. 2019. "Cerebral Cortical Folding, Parcellation, and Connectivity in Humans, Nonhuman Primates, and Mice." *Proc Natl Acad Sci USA* 116: 26173–80. D. C. Van Essen, C. J. Donahue, and M. F. Glasser. 2018. "Development and Evolution of Cerebral and Cerebellar Cortex." *Brain Behav Evol* 91: 158–69.

4. M. L. Li, H. Tang, Y. Shao, M. S. Wang, H. B. Xu, S. Wang, D. M. Irwin, et al. 2020. "Evolution and Transition of Expression Trajectory during Human Brain Development." *BMC Evol Biol* 20: 72.

5. S. Benito-Kwiecinski, S. L. Giandomenico, M. Sutcliffe, E. S. Riis, P. Freire-Pritchett, I. Kelava, S. Wunderlich, et al. 2021. "An Early Cell Shape Transition Drives Evolutionary Expansion of the Human Forebrain." *Cell* 184: 2084–102.

6. A. Levchenko, A. Kanapin, A. Samsonova, and R. R. Gainetdinov. 2018. "Human Accelerated Regions and Other Human-Specific Sequence Variations in the Context of Evolution and Their Relevance for Brain Development." *Genome Biol Evol* 10: 166–88.

7. C. A. Trujillo, E. S. Rice, N. K. Schaefer, I. A. Chaim, E. C. Wheeler, A. A. Madrigal, J. Buchanan, et al. 2021. "Reintroduction of the Archaic Variant of *NOVA1* in Cortical Organoids Alters Neurodevelopment." *Science* 371: eaax2537.

8. P. Broca. 1861. "Perte de la Parole: Ramollissement chronique et destruction partielle du lobe antérieur gauche du cerveau." *Bulletin de la Société Anthropologique* 2: 235–38. Translated in E. A. Berker, A. H. Berker, and A. Smith. 1986. "Localization of Speech in the Third Left Frontal Convolution." *Arch Neurol* 43: 1065–72.

9. P. Broca. 1861. "Nouvelle Observation d'Aphémie Produite par une Lésion de la Moitié Postérieure des Deuxième et Troisième Circonvolution Frontales Gauches." *Bull Soc Anat* 36: 398–407.

10. C. Wernicke. 1874. *Der Aphasische Symptomencomplex: Eine Psychologische Studie auf Anatomischer Basis.* Breslau: Max Cohn & Weigert.

11. J. Li, D. E. Osher, H. A. Hansen, and Z. M. Saygin. 2020. "Innate Connectivity Patterns Drive the Development of the Visual Word Form Area." *Sci Rep* 10: 18039. C. S. Lai, S. E. Fisher, J. A. Hurst, F. Vargha-Khadem, and A. P. Monaco. 2001. "A Forkhead-Domain Gene Is Mutated in a Severe Speech and Language Disorder." *Nature* 413: 519–23.

12. M. Co, A. G. Anderson, and G. Konopka. 2020. "FOXP Transcription Factors in Vertebrate Brain Development, Function, and Disorders." *Wiley Interdiscip Rev Dev Biol* 9: e375.

13. W. Enard, S. Gehre, K. Hammerschmidt, S. M. Hölter, T. Blass, M. Somel, M. K. Brückner, et al. 2009. "A Humanized Version of Foxp2 Affects Cortico-Basal Ganglia Circuits in Mice." *Cell* 137: 961–71.

14. J. den Hoed and S. E. Fisher. 2020. "Genetic Pathways Involved in Human Speech Disorders." *Curr Opin Genet Dev* 65: 103–11.

15. J. den Hoed and S. E. Fisher. 2020. "Genetic Pathways Involved in Human Speech Disorders." *Curr Opin Genet Dev* 65: 103–11.

16. S. Curtiss. 1977. *Genie: A Psycholinguistic Study of a Modern-Day "Wild Child."* Perspectives in Neurolinguistics and Psycholinguistics. Cambridge, MA: Academic Press.

17. V. Duboc, P. Dufourcq, P. Blader, and M. Roussigné. 2015. "Asymmetry of the Brain: Development and Implications." *Annu Rev Genet* 49: 647–72.

18. R. Sperry. 1982. "Some Effects of Disconnecting the Cerebral Hemispheres." Nobel Lecture, *Biosci Rep* 2: 265–76.

19. S. C. Harvey and H. E. Orbidans. 2011. "All Eggs Are Not Equal: The Maternal Environment Affects Progeny Reproduction and Developmental Fate in *Caenorhabditis elegans*. *PLoS One* 6: e25840.

20. L. H. Lumey, A. D. Stein, H. S. Kahn, K. M. van der Pal-de Bruin, G. J. Blauw, P. A. Zybert, and E. S. Susser. 2007. "Cohort Profile: The Dutch Hunger Winter Families Study." *Int J Epidemiol* 36: 1196–204.

21. J. P. Curley and F. A. Champagne. 2016. "Influence of Maternal Care on the Developing Brain: Mechanisms, Temporal Dynamics and Sensitive Periods." *Front Neuroendocrinol* 40: 52–66. R. Feldman, K. Braun, and F. A. Champagne. 2019. "The Neural Mechanisms and Consequences of Paternal Caregiving." *Nat Rev Neurosci* 20: 205–24.

22. B. G. Dias and K. J. Ressler. 2014. "Parental Olfactory Experience Influences Behavior and Neural Structure in Subsequent Generations." *Nat Neurosci* 17: 89–96.

23. R. Toro, J. B. Poline, G. Huguet, E. Loth, V. Frouin, T. Banaschewski, G. J. Barker, et al. 2015. "Genomic Architecture of Human Neuroanatomical Diversity." *Mol Psychiatr* 20: 1011–16.

24. D. Duan, S. Xia, I. Rekik, Z. Wu, L. Wang, W. Lin, J. H. Gilmore, D. Shen, and G. Li. 2020. "Individual Identification and Individual Variability Analysis Based on Cortical Folding Features in Developing Infant Singletons and Twins." *Hum Brain Mapp* 41: 1985–2003.

25. T. T. Brown. 2017. "Individual Differences in Human Brain Development." *Wiley Interdiscip Rev Cogn Sci* 8: e1389. A. I. Becht and K. L. Mills. 2020. "Modeling Individual Differences in Brain Development." *Biol Psychiatry* 88: 63–69.

26. B. Hagekull and G. Bohlin. 2003. "Early Temperament and Attachment as Predictors of the Five Factor Model of Personality." *Attach Hum Dev* 5: 2–18.

27. C. M. Kelsey, K. Farris, and T. Grossmann. 2021. "Variability in Infants' Functional Brain Network Connectivity Is Associated with Differences in Affect and Behavior." *Front Psychiatry* 12: 685754.

28. K. L. Jang, W. J. Livesley, and P. A. Vernon. 1996. "Heritability of the Big Five Personality Dimensions and Their Facets: A Twin Study." *J Pers* 64: 577–91. S. Sanchez-Roige, J. C. Gray, J. MacKillop, C. H. Chen, and A. A. Palmer. 2018. "The Genetics of Human Personality." *Genes Brain Behav* 17: e12439.

29. E. Castaldi, C. Lunghi, and M. C. Morrone. 2020. "Neuroplasticity in Adult Human Visual Cortex." *Neurosci Biobehav Rev* 112: 542–52. I. Fine and J. M. Park. 2018. "Blindness and Human Brain Plasticity." *Annu Rev Vis Sci* 4: 337–56.

30. M. C. Diamond, D. Krech, and M. R. Rosenzweig. 1964. "The Effects of an Enriched Environment on the Histology of the Rat Cerebral Cortex." *J Comp Neurol* 123: 111–20.

31. Ö. de Manzano and F. Ullén. 2018. "Same Genes, Different Brains: Neuroanatomical Differences between Monozygotic Twins Discordant for Musical Training." *Cereb Cortex* 28: 387–94.

INDEX

Page numbers indicated with italics represent figures.